# The
# Physics Formulary

## for bilingual teaching

# The
# Physics Formulary
## for bilingual teaching

***Author***: Matthias Heidrich, PhD

***Content advisor***: Daniel Schläpfer, PhD

***Language advisor***: Adam Stewart, BSc

**Matthias Heidrich**, PhD, is the author of several reference books on physics and mathematics. He studied physics and mathematics at the University of Heidelberg. After five research years at CERN in Geneva, he joined *Swisslife* in Zurich in 1997 to dedicate himself to insurance mathematics. In 2002, Matthias Heidrich became a teacher for physics and mathematics at AKAD in Zurich. Since 2004, he has taught physics and related subjects at Kantonsschule Wil (Switzerland) in German and English language.

**Daniel Schläpfer**, PhD, studied geography and physics at the University of Zurich. Since 2002, he has been a physics teacher at Kantonsschule Wil. Moreover, he is involved in remote sensing and imaging spectroscopy research in collaboration with universities, institutions and companies worldwide and lectures at the University of Zurich.

**Adam Stewart**, BSc, studied ecology and conservation at Lancaster University (United Kingdom). Also a dedicated linguist, he has been the English language assistant at Kantonsschule Wil since 2018.

Impressum:

© 2020 Matthias Heidrich

Herstellung und Verlag: BoD - Books on Demand, Norderstedt

ISBN: 978-3-75193-034-5

Bibliografische Information der Deutschen

Nationalbibliothek

# Contents

# I.   Introduction

This physics formulary is designed for use in immersive or bilingual teaching in secondary and high schools in German-speaking areas. All technical terms have been diligently researched, with special care being taken to ensure that the correct British English expressions have been given.  In the index at the end of the book, a German translation is provided for all technical terms listed.

As physics curricula vary not only from region to region but also from school to school, this book contains formulas and tables covering all possible aspects of the subject.

It may also prove useful to students and teachers at university level in both, German- and English-speaking areas.

# II.   Physics Formulas

## 1. Kinematics

| Time | $t$ | s |
| --- | --- | --- |
| Position | $s$ | m |
| Position vector | $\vec{s}\ or\ \vec{r}$ | m |
| Displacement | $\Delta\vec{s}\ or\ \Delta\vec{r}$ | m |

### Definitions: Speed, velocity and acceleration

| (Instantaneous) speed | $v$ | $m \cdot s^{-1}$ |
| --- | --- | --- |
| (Instantaneous) velocity | $\vec{v}$ | $m \cdot s^{-1}$ |
| (Instantaneous) acceleration | $a$ | $m \cdot s^{-2}$ |
| (Instantaneous) acceleration vector | $\vec{a}$ | $m \cdot s^{-2}$ |
| Average speed | $\langle v \rangle = \dfrac{\Delta s}{\Delta t} = \dfrac{s_2 - s_1}{t_2 - t_1}$ | $m \cdot s^{-1}$ |
| Average velocity | $\langle \vec{v} \rangle = \dfrac{\Delta \vec{s}}{\Delta t} = \dfrac{\vec{s}_2 - \vec{s}_1}{t_2 - t_1}$ | $m \cdot s^{-1}$ |
| Average acceleration | $\langle a \rangle = \dfrac{\Delta v}{\Delta t} = \dfrac{v_2 - v_1}{t_2 - t_1}$ | $m \cdot s^{-2}$ |
| Average acceleration vector | $\langle \vec{a} \rangle = \dfrac{\Delta \vec{v}}{\Delta t} = \dfrac{\vec{v}_2 - \vec{v}_1}{t_2 - t_1}$ | $m \cdot s^{-2}$ |

### Uniform motion with starting point

| Speed | $v = const.$ | $m \cdot s^{-1}$ |
| --- | --- | --- |
| Starting point | $s_0$ | m |
| Position | $s = s_0 + v \cdot t$ | m |
| Average speed | $\langle v \rangle = v$ | $m \cdot s^{-1}$ |

## Uniformly accelerated motion

| | | |
|---|---|---|
| Acceleration | $a = const.$ | $m \cdot s^{-2}$ |
| Speed | $v = a \cdot t$ | $m \cdot s^{-1}$ |
| Speed | $v = \sqrt{2 \cdot a \cdot s}$ | $m \cdot s^{-1}$ |
| Position | $s = \dfrac{1}{2} \cdot v \cdot t$ | $m$ |
| Position | $s = \dfrac{1}{2} \cdot a \cdot t^2$ | $m$ |
| Average speed | $\langle v \rangle = \dfrac{v}{2}$ | $m \cdot s^{-1}$ |

## Uniformly accelerated motion with initial speed and starting point

| | | |
|---|---|---|
| Acceleration | $a = const.$ | $m \cdot s^{-2}$ |
| Starting point | $s_0$ | $m$ |
| Initial speed | $v_0$ | $m \cdot s^{-1}$ |
| Speed | $v = v_0 + a \cdot t$ | $m \cdot s^{-1}$ |
| Position | $s = s_0 + v_0 \cdot t + \dfrac{1}{2} \cdot a \cdot t^2$ | $m$ |
| Speed | $v = \sqrt{2 \cdot a \cdot (s - s_0) + v_0^2}$ | $m \cdot s^{-1}$ |

## Braking distance and braking time

| | | |
|---|---|---|
| Acceleration (deceleration) | $a = const.\,; a < 0$ | $m \cdot s^{-2}$ |
| Initial speed | $v_0$ | $m \cdot s^{-1}$ |
| Braking distance | $s_B = -\dfrac{v_0{}^2}{2 \cdot a}$ | $m$ |
| Braking time | $t_B = -\dfrac{v_0}{a}$ | $s$ |

## Vertical throw

| | | |
|---|---|---|
| Acceleration (deceleration) | $a = -g = const.$ | $m \cdot s^{-2}$ |
| Initial speed | $v_0$ | $m \cdot s^{-1}$ |
| Position, height | $y = v_0 \cdot t - \dfrac{1}{2} \cdot g \cdot t^2$ | $m$ |
| Speed | $v = v_0 - g \cdot t$ | $m \cdot s^{-1}$ |
| Rising time | $t_h = \dfrac{v_0}{g}$ | $s$ |
| Flight time | $t_d = \dfrac{2 \cdot v_0}{g}$ | $s$ |
| Maximum height | $y_h = \dfrac{v_0^2}{2 \cdot g}$ | $m$ |

## Horizontal throw

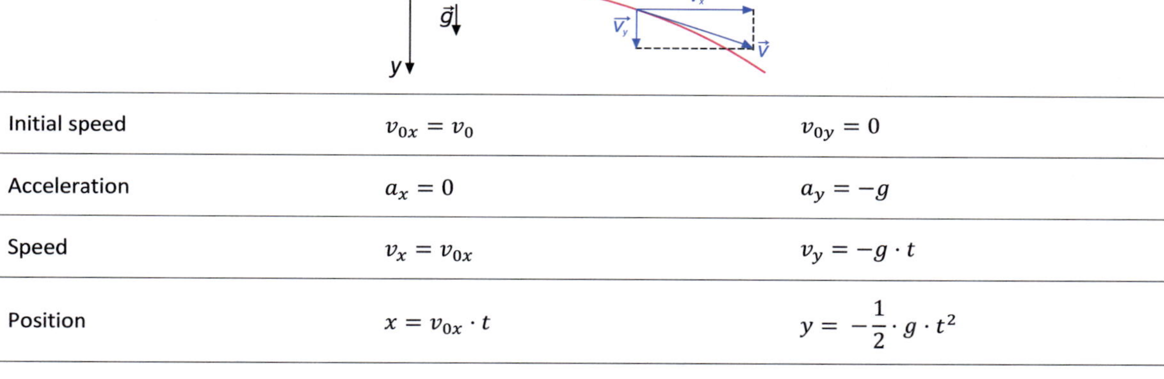

| | | |
|---|---|---|
| Initial speed | $v_{0x} = v_0$ | $v_{0y} = 0$ |
| Acceleration | $a_x = 0$ | $a_y = -g$ |
| Speed | $v_x = v_{0x}$ | $v_y = -g \cdot t$ |
| Position | $x = v_{0x} \cdot t$ | $y = -\dfrac{1}{2} \cdot g \cdot t^2$ |
| Path | | $y = -\left(\dfrac{g}{2 \cdot v_{0x}^2}\right) \cdot x^2$ |

## Projectile motion

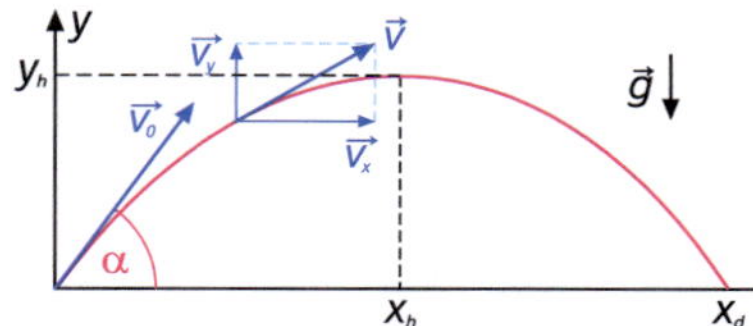

| | | |
|---|---|---|
| Initial speed | $v_{0x} = v_0 \cdot \cos \alpha$ | $v_{0y} = v_0 \cdot \sin \alpha$ |
| Acceleration | $a_x = 0$ | $a_y = -g$ |

| Speed | $v_x = v_{0x}$ | $v_y = v_{0y} - g \cdot t$ |
|---|---|---|
| Position | $x = v_{0x} \cdot t$ | $y = v_{0y} \cdot t - \frac{1}{2} \cdot g \cdot t^2$ |

| Path | $y = \dfrac{v_{0y}}{v_{0x}} \cdot x - \dfrac{g}{2 \cdot v_{0x}^2} \cdot x^2$ |
|---|---|
| | $y = x \cdot \tan\alpha - \dfrac{g}{2 \cdot v_0^2 \cdot \cos^2\alpha} \cdot x^2$ |

| Throwing distance | $x_d = \dfrac{2 \cdot v_{0x} v_{0y}}{g}$ |
|---|---|
| Flight time | $t_d = \dfrac{2 \cdot v_{0y}}{g}$ |
| Maximum height | $y_h = \dfrac{v_{0y}^2}{2 \cdot g}$ |
| Point of maximum height | $x_h = \dfrac{x_d}{2} = \dfrac{v_{0x} v_{0y}}{g}$ |
| Rising time | $t_h = \dfrac{v_{0y}}{g}$ |

## Uniform circular motion

| Speed | $v = const.$ | | $\mathrm{m \cdot s^{-1}}$ |
|---|---|---|---|
| Radius | $r$ | | $\mathrm{m}$ |
| Length of arc | $s$ | | $\mathrm{m}$ |
| Time period | $T$ | | $\mathrm{s}$ |
| Frequency | $f = \dfrac{1}{T}$ | | $\mathrm{Hz}$ |
| Angle | $\Delta\varphi = \dfrac{s}{r}$ | 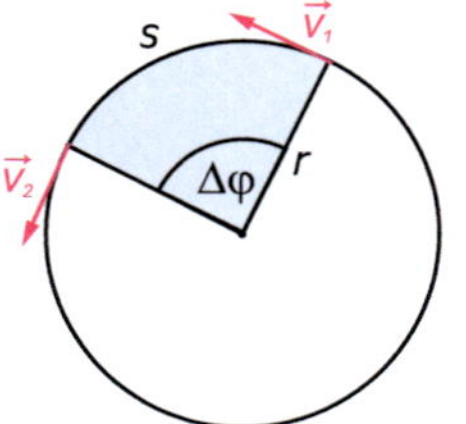 | $\mathrm{rad}$ |
| Angular speed | $\omega = \dfrac{\Delta\varphi}{\Delta t} = \dfrac{2 \cdot \pi}{T} = 2 \cdot \pi \cdot f$ | | $\mathrm{s^{-1}}$ |
| Speed | $v = \omega \cdot r$ | | $\mathrm{m \cdot s^{-1}}$ |

| Centripetal acceleration vector | $\vec{a}_{CP} = -\omega^2 \cdot \vec{r}$ | 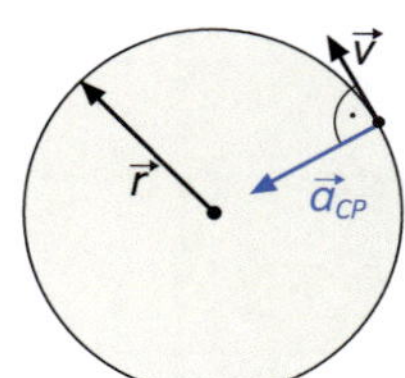 | $m \cdot s^{-2}$ |
| --- | --- | --- | --- |
| Centripetal acceleration | $a_{CP} = \dfrac{v^2}{r} = \omega^2 \cdot r$ | | $m \cdot s^{-2}$ |

# 2. Dynamics

| Mass | $m$ | kg |
| --- | --- | --- |
| Velocity | $\vec{v}$ | $m \cdot s^{-1}$ |
| Force vector | $\vec{F}$ | N |

## Properties of materials

| Density | $\rho = \dfrac{m}{V}$ | | $kg \cdot m^{-3}$ |
| --- | --- | --- | --- |
| Tensile stress in a wire | $\sigma = \dfrac{F_T}{A}$ | $F_T$: *tension*<br>$A$: *wire cross section* | $N \cdot m^{-2}$ |
| Tensile strain of a wire | $\varepsilon = \dfrac{\Delta \ell}{\ell}$ | $\dfrac{\Delta \ell}{\ell}$: *relative change in length* | – |
| Modulus of elasticity | $E = \dfrac{\sigma}{\varepsilon}$ | $\sigma$: *tensile stress*<br>$\varepsilon$: *tensile strain* | $N \cdot m^{-2}$ |

## Newtonian laws and Hooke's law

| Momentum | $\vec{p} = m \cdot \vec{v}$ | | N · s |
| --- | --- | --- | --- |
| Impulse | $\Delta \vec{p} = \vec{F} \cdot \Delta t$ | | N · s |
| 1st Newtonian law | $\vec{F} = \vec{o} \Leftrightarrow \vec{v} = const.$ | | |
| Force,<br>2nd Newtonian law | $\vec{F} = \dfrac{\Delta \vec{p}}{\Delta t}$ | | N |
| | $\vec{F} = m \cdot \vec{a}$ | $if\ m = const.$ | N |
| | $\vec{F} = \dfrac{\Delta m}{\Delta t} \cdot \vec{v}$ | $if\ \vec{v} = const.$ | N |

| | | | |
|---|---|---|---|
| 3rd Newtonian law (action=reaction) | $\vec{F}_1 = -\vec{F}_2$ | $\vec{F}_1$: *force* <br> $\vec{F}_2$: *counterforce* | |
| Weight, gravitational force | $\vec{F}_G = m \cdot \vec{g}$ | $g$: *acceleration due to gravity* | N |
| Spring force vector, Hooke's law | $\vec{F}_S = -k \cdot \vec{y}$ | $k$: *spring constant* <br> $y$: *spring elongation* <br> $\vec{F}_P$: *pulling force* | N |
| Spring force, Hooke's law | $F_S = k \cdot y$ | $k$: *spring constant* <br> $y$: *spring elongation* | N |

## Normal force and friction

| | | | |
|---|---|---|---|
| Normal force vector | $\vec{F}_N = -\vec{F}_G$ | *level plane* | N |
| Normal force | $F_N = F_G$ | *level plane* | N |
| Normal force vector | $\vec{F}_N = -\vec{F}_\perp$ | $(\rightarrow)$*inclined plane* | N |
| Normal force | $F_N = F_G \cdot \cos \alpha$ | $(\rightarrow)$*inclined plane* | N |
| Static friction force | $F_{FS} \leq F_{FS,max} = \mu_S \cdot F_N$ | $\mu_S$: *static friction coefficient* | N |
| Dynamic friction force | $F_{FD} = \mu_D \cdot F_N$ | $\mu_D$: *dynamic friction coefficient* | N |
| Rolling friction force | $F_{FR} = \mu_R \cdot F_N$ | $\mu_R$: *rolling friction coefficient* | N |
| Stokes' law (laminar flow) | $F_L = 6 \cdot \pi \cdot r \cdot \eta \cdot v$ | $r$: *ball radius* <br> $\eta$: *viscosity* <br> $v$: *ball speed* | N |
| Fluid resistance, drag (turbulent flow) | $F_D = \dfrac{1}{2} \cdot c_D \cdot \rho \cdot A \cdot v^2$ | $c_D$: *drag coefficient* <br> $v$: *object speed* <br> $\rho$: *density of the fluid* <br> $A$: *cross sectional area of object* | N |

## Uniform circular motion

| | | | |
|---|---|---|---|
| Centripetal force | $$F_{CP} = m \cdot a_{CP} = m \cdot \omega^2 \cdot r = m \cdot \frac{v^2}{r}$$ | 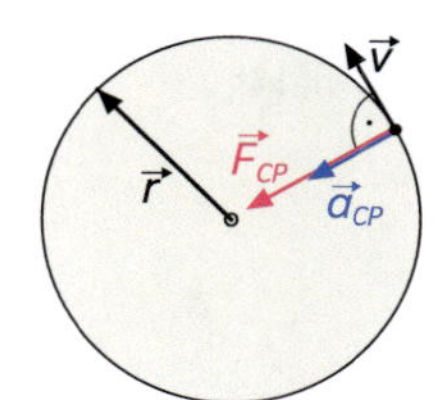 | N |

| | | | |
|---|---|---|---|
| Centripetal force vector | $$\vec{F}_{CP} = m \cdot \vec{a}_{CP} = -m \cdot \omega^2 \cdot \vec{r}$$ | | N |

## Technical applications

### *Hoist*

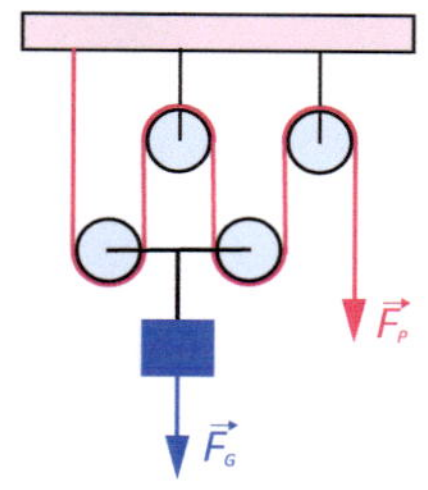

| | | | |
|---|---|---|---|
| $$F_P = \frac{F_G}{n}$$ | *n: number of pulleys* | N |
| $$s = n \cdot h$$ | *h: lifting height*<br>*s: pulling distance* | m |

### *Lever and beam balance*

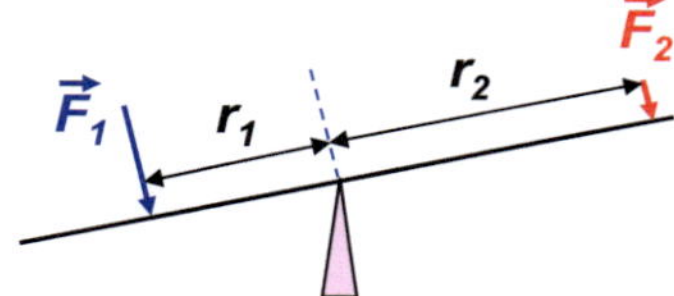

| | | |
|---|---|---|
| $$M = r \cdot F$$ | *M: torque*<br>*r: lever arm* | Nm |
| $$r_1 \cdot F_1 = r_2 \cdot F_2$$<br>$$M_1 = M_2$$ | *equilibrium condition* | |

### *Inclined plane*

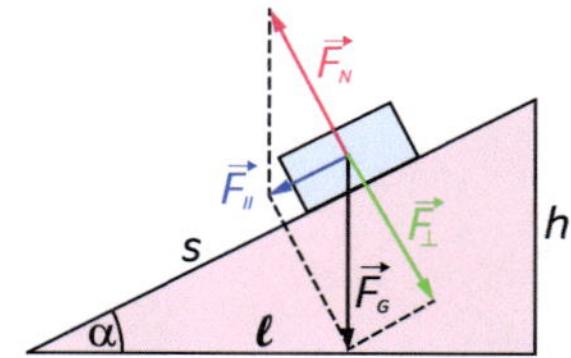

| | | |
|---|---|---|
| $$p = \tan \alpha = \frac{h}{\ell}$$ | *p: slope in percent*<br>*α: slope angle*<br>*h: height of the plane*<br>*ℓ: base of the plane* | – |
| $$\sin \alpha = \frac{h}{s}$$ | *s: length of the slope* | – |
| $$F_\parallel = F_G \cdot \sin \alpha$$ | *F$_\parallel$: force ∥ plane* | N |
| $$F_\perp = F_G \cdot \cos \alpha$$ | *F$_\perp$: force ⊥ plane* | N |
| $$\vec{F}_N = -\vec{F}_\perp$$ | | N |
| $$F_N = F_G \cdot \cos \alpha$$ | | N |

## 3. Gravitation

| | | | |
|---|---|---|---|
| Gravitational force | $F_G = G \cdot \dfrac{m_1 \cdot m_2}{r^2}$ | $G$: gravitational constant<br>$m_1$: mass of object 1<br>$m_2$: mass of object 2<br>$r$: distance | N |
| Gravitational force vector | $\vec{F}_G = -G \cdot \dfrac{m_1 \cdot m_2}{r^2} \cdot \dfrac{\vec{r}}{r}$ | $G$: gravitational constant<br>$m_1$: mass of object 1<br>$m_2$: mass of object 2<br>$r$: distance<br>$\vec{r}$: displacement | N |
| 3rd Kepler law (by Kepler) | $\dfrac{r^3}{T^2} = const.$ | $r$: orbital radius of planet<br>$T$: orbit period of planet | |
| 3rd Kepler law (by Newton) | $\dfrac{r^3}{T^2} = \dfrac{G \cdot m_1}{4 \cdot \pi^2}$ | $G$: gravitational constant<br>$m_1$: central mass<br>$r$: orbital radius<br>$T$: orbit period | |

## 4. Work, Power and Energy

| | | |
|---|---|---|
| Mass of a certain object | $m$ | kg |
| Displacement of a certain object | $\vec{s}$ | m |
| Velocity of a certain object | $\vec{v}$ | $m \cdot s^{-1}$ |
| Force exerted on a certain object | $\vec{F}$ | N |

### Work

| | | | |
|---|---|---|---|
| Work | $W = F \cdot s$ | if $\vec{F} \parallel \vec{s}$ | J |
| Work | $W = F \cdot s \cdot \cos \alpha$ | $\alpha$: angle between $\vec{F}$ and $\vec{s}$ | J |
| Lifting work | $W_L = m \cdot g \cdot h$ | $g$: acceleration due to gravity<br>$h$: lifting height | J |
| Acceleration work | $W_A = m \cdot a \cdot s$ | $a$: acceleration of object | J |
| Acceleration work | $W_A = \dfrac{1}{2} \cdot m \cdot v^2$ | $v$: final speed of object | J |

| | | | |
|---|---|---|---|
| Friction work | $W_F = \mu \cdot F_N \cdot s$ | $\mu$: friction coefficient<br>$F_N$: normal force | J |
| Friction work (level plane) | $W_F = \mu \cdot m \cdot g \cdot s$ | $\mu$: friction coefficient<br>$g$: acceleration due to gravity | J |
| Work performed on a spring | $W_S = \frac{1}{2} \cdot k \cdot y^2$ | $k$: spring constant<br>$y$: spring elongation | J |

## Power

| | | | |
|---|---|---|---|
| Power | $P = \dfrac{W}{t}$ | $W$: work performed<br>$t$: time taken | W |
| Power | $P = F \cdot v \cdot \cos \alpha$ | $\alpha$: angle between $\vec{F}$ and $\vec{v}$ | W |
| Power | $P = F \cdot v$ | if $\vec{F} \parallel \vec{v}$ | W |

## Energy

| | | | |
|---|---|---|---|
| Energy | $E$ | | J |
| Potential energy (PE) | $E_P = m \cdot g \cdot h$ | $g$: acceleration due to gravity<br>$h$: object height | J |
| Kinetic energy (KE) | $E_K = \frac{1}{2} \cdot m \cdot v^2$ | | J |
| Spring energy | $E_S = \frac{1}{2} \cdot k \cdot y^2$ | $k$: spring constant<br>$y$: spring elongation | J |
| Total energy | $E_{tot} = E_P + E_K + E_S + \cdots$ | | J |
| Energy conservation law (open system) | $E_A = E_B + W$<br>$E_A - E_B = \Delta E = W$ | $E_B$: total energy before a change<br>$E_A$: total energy after a change<br>$W$: work performed on ($W > 0$) or by ($W < 0$) the system | J |
| Energy conservation law (closed system) | $E_A = E_B$ | $E_B$: total energy before a change<br>$E_A$: total energy after a change | J |
| Efficiency | $\eta = \dfrac{E_{out}}{E_{in}}$ | $E_{out}$: useful energy output<br>$E_{in}$: total energy input | – |

| Efficiency | $\eta = \dfrac{P_{out}}{P_{in}}$ | $P_{out}$: useful power output<br>$P_{in}$: total power input | – |

# 5. Hydrostatics

| Pressure | $p = \dfrac{F}{A}$ | $F$: force<br>$A$: area | Pa |
| --- | --- | --- | --- |
| Hydrostatic pressure | $p = \rho \cdot g \cdot h$ | $\rho$: density of the liquid<br>$g$: acceleration due to gravity<br>$h$: immersion depth | Pa |
| Total pressure in a liquid | $p = p_0 + \rho \cdot g \cdot h$ | $p_0$: air pressure on the liquid<br>$\rho$: density of a liquid<br>$g$: acceleration due to gravity<br>$h$: immersion depth | Pa |
| Buoyancy | $F_B = \rho_F \cdot g \cdot V = m_F \cdot g$ | $\rho_F$: density of the fluid<br>$g$: acceleration due to gravity<br>$V$: volume<br>$m_F$: mass of the displaced fluid | N |
| Boyle-Mariotte's law | $\vartheta = const. \Rightarrow p \cdot V = const.$ | $\vartheta$: gas temperature $[°C]$<br>$p$: gas pressure<br>$V$: gas volume | |
| Boyle-Mariotte's law | $\vartheta = const. \Rightarrow \dfrac{p}{\rho} = const.$ | $\vartheta$: gas temperature $[°C]$<br>$p$: gas pressure<br>$\rho$: gas density | |
| Barometric formula (pressure) | $p(h) = p(0) \cdot exp\left(-\dfrac{\rho_S}{p_S} \cdot g_S \cdot h\right)$ | $p(h)$: air pressure at height $h$<br>$p(0)$: air pressure at height 0<br>$\rho_S$: air density in standard conditions<br>$p_S$: standard pressure<br>$g_S$: standard acceleration due to gravity | Pa |
| Barometric formula (density) | $\rho(h) = \rho(0) \cdot exp\left(-\dfrac{\rho_S}{p_S} \cdot g_S \cdot h\right)$ | $\rho(h)$: air density at height $h$<br>$\rho(0)$: air density at height 0<br>$\rho_S$: air density in standard conditions<br>$p_S$: standard pressure<br>$g_S$: standard acceleration due to gravity | $kg \cdot m^{-3}$ |

# 6. Heat

| Quantity | Formula | Description | Unit |
|---|---|---|---|
| Celsius temperature | $\vartheta$ | | °C |
| Absolute temperature | $T$ | | K |
| Heat | $Q$ | | J |
| Internal energy | $U$ | | J |
| Thermal expansion of solids (linear) | $\dfrac{\Delta\ell}{\ell_0} = \alpha \cdot \Delta\vartheta$ | $\ell_0$: *initial length* <br> $\ell$: *final length* <br> $\Delta\ell = \ell - \ell_0$ <br> $\Delta\vartheta = temperature\ change$ <br> $\alpha$: *(linear) thermal expansion coefficient* | – |
| Thermal expansion of solids (linear) | $\ell = \ell_0(1 + \alpha \cdot \Delta\vartheta)$ | $\ell_0$: *initial length* <br> $\ell$: *final length* <br> $\Delta\vartheta = temperature\ change$ <br> $\alpha$: *(linear) thermal expansion coefficient* | m |
| Thermal expansion of solids (volumetric) | $\dfrac{\Delta V}{V_0} = \gamma \cdot \Delta\vartheta$ | $V_0$: *initial volume* <br> $V$: *final volume* <br> $\Delta V = V - V_0$ <br> $\Delta\vartheta = temperature\ change$ <br> $\gamma \approx 3 \cdot \alpha$: *(volumetric) thermal expansion coefficient* | – |
| Thermal expansion of solids (volumetric) | $V = V_0(1 + \gamma \cdot \Delta\vartheta)$ | $V_0$: *initial volume* <br> $V$: *final volume* <br> $\Delta\vartheta = temperature\ change$ <br> $\gamma \approx 3 \cdot \alpha$: *(volumetric) thermal expansion coefficient* | $m^3$ |
| Thermal expansion of liquids (volumetric) | $\dfrac{\Delta V}{V_0} = \gamma \cdot \Delta\vartheta$ | $V_0$: *initial volume* <br> $V$: *final volume* <br> $\Delta V = V - V_0$ <br> $\Delta\vartheta = temperature\ change$ <br> $\gamma$: *(volumetric) thermal expansion coefficient* | – |
| Thermal expansion of liquids (volumetric) | $V = V_0(1 + \gamma \cdot \Delta\vartheta)$ | $V_0$: *initial volume* <br> $V$: *final volume* <br> $\Delta\vartheta = temperature\ change$ <br> $\gamma$: *(volumetric) thermal expansion coefficient* | $m^3$ |
| Compressibility of a fluid | $\chi = \dfrac{1}{\Delta p} \cdot \dfrac{\Delta V}{V}$ | $\dfrac{\Delta V}{V}$: *relative change in volume* <br> $\Delta p$: *change in pressure* | $Pa^{-1}$ |
| Boyle-Mariotte's law | $T = const. \Rightarrow p \cdot V = const.$ | $T$: *gas temperature* <br> $p$: *gas pressure* <br> $V$: *gas volume* | |

| | | |
|---|---|---|
| Gay-Lussac's law | $p = const. \Rightarrow \dfrac{V}{T} = const.$ | $T$: gas temperature<br>$p$: gas pressure<br>$V$: gas volume |
| Amonton's law | $V = const. \Rightarrow \dfrac{p}{T} = const.$ | $T$: gas temperature<br>$p$: gas pressure<br>$V$: gas volume |
| Ideal gas equation | $p \cdot V = n \cdot R \cdot T$ | $p$: gas pressure<br>$V$: gas volume<br>$n$: amount of substance<br>$R$: gas constant<br>$T$: gas temperature |
| Average kinetic energy of a gas particle | $\langle E_K \rangle = \dfrac{1}{2} \cdot m \cdot \langle v^2 \rangle$ | $m$: mass of a gas particle<br>$v$: speed of a gas particle    J |
| Average kinetic energy of a gas particle | $\langle E_K \rangle = \dfrac{3}{2} \cdot k \cdot T$ | $k$: Boltzmann constant<br>$T$: gas temperature    J |
| 1st law of thermodynamics | $U_A = U_B + Q + W$<br><br>or<br><br>$\Delta U = Q + W$ | $U_A$: internal energy after a change<br>$U_B$: internal energy before a change<br>$Q$: heat received<br>$W$: work performed on ($W > 0$) or by ($W < 0$) the system<br>$\Delta U = U_A - U_B$: change in internal energy |
| Work in a cylinder | $W = p \cdot \Delta V$ | $p$: mean cylinder pressure<br>$\Delta V$: cylinder capacity    J |
| Power in a cylinder | $P = p \cdot \Delta V \cdot f$ | $p$: mean cylinder pressure<br>$\Delta V$: cylinder capacity<br>$f$: engine frequency    W |
| Maximum efficiency | $\eta_{max} = \dfrac{\Delta T}{T_{max}}$ | $\Delta T$: temperature difference<br>$T_{max}$: maximum higher temperature    – |
| Maximum efficiency | $\eta_{max} = \dfrac{Q_{in} - Q_{out}}{Q_{in}}$ | $Q_{in}$: heat input<br>$Q_{out}$: heat output    – |
| Specific heat capacity | $c = \dfrac{Q}{m \cdot \Delta\vartheta}$ | $Q$: heat received<br>$m$: mass<br>$\Delta\vartheta$: temperature change    $\dfrac{\text{J}}{\text{kg} \cdot \text{K}}$ |
| Adiabatic coefficient (gases) | $\kappa = \dfrac{c_p}{c_v}$ | $c_p$: specific heat capacity if $p = const.$<br>$c_v$: specific heat capacity if $V = const.$    – |

| | | | |
|---|---|---|---|
| Temperature of a mixture | $\vartheta_m = \frac{c_1 \cdot m_1 \cdot \vartheta_1 + c_2 \cdot m_2 \cdot \vartheta_2}{c_1 \cdot m_1 + c_2 \cdot m_2}$ | $c_1$: *c of object 1*<br>$c_2$: *c of object 2*<br>$m_1$: *mass of object 1*<br>$m_1$: *mass of object 2*<br>$\vartheta_1$: *temperature of object 1*<br>$\vartheta_2$: *temperature of object 2* | °C |
| Specific latent heat of fusion | $L_f = \frac{Q}{m}$ | $Q$: *heat received*<br>$m$: *mass of the substance* | $\mathrm{J \cdot kg^{-1}}$ |
| Specific latent heat of vaporization | $L_v = \frac{Q}{m}$ | $Q$: *heat received*<br>$m$: *mass of the substance* | $\mathrm{J \cdot kg^{-1}}$ |
| Heat conduction | $\frac{\Delta Q}{\Delta t} = \lambda \cdot A \cdot \frac{\Delta T}{d}$<br>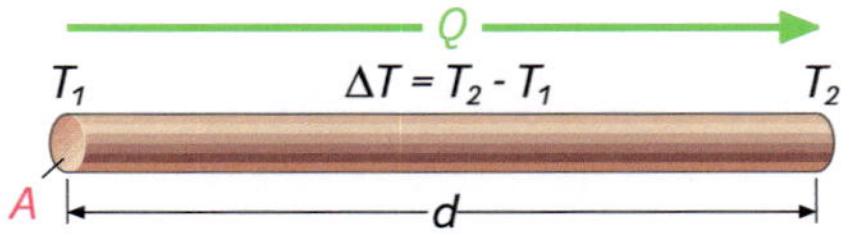<br> | $\frac{\Delta Q}{\Delta t}$: *heat conducted per time*<br>$\lambda$: *heat conduction coefficient*<br>$A$: *cross section*<br>$\Delta T$: *temperature difference*<br>$d$: *thickness* | W |
| Heat transfer | $\frac{\Delta Q}{\Delta t} = U \cdot A \cdot \Delta T$ | $\frac{\Delta Q}{\Delta t}$: *heat conducted per time*<br>$U$: *heat transfer coefficient*<br>$A$: *cross section*<br>$\Delta T$: *temperature difference* | W |
| Stefan-Boltzmann law (thermal radiation) | $\frac{\Delta Q}{\Delta t} = \varepsilon \cdot \sigma \cdot A \cdot T^4$ | $\frac{\Delta Q}{\Delta t}$: *heat radiated per time*<br>$\varepsilon$: *emissivity*<br>$\sigma$: *Stefan – Boltzmann constant*<br>$A$: *surface area*<br>$T$: *temperature* | W |
| Heating value | $H_L = \frac{Q}{m}$ | $Q$: *heat released during combustion*<br>$m$: *mass of the substance* | $\mathrm{J \cdot kg^{-1}}$ |

## 7. Oscillations

| | | |
|---|---|---|
| Time | $t$ | s |
| Amplitude | $\hat{y}$ | m |
| Phase | $\varphi_0$ | rad |
| Period | $T$ | s |
| Frequency | $f = \frac{1}{T}$ | Hz |
| Angular speed | $\omega = \frac{2 \cdot \pi}{T} = 2 \cdot \pi \cdot f$ | $\mathrm{s^{-1}}$ |

## Harmonic oscillations

| | | |
|---|---|---|
| Elongation | $y = \hat{y} \cdot cos(\omega \cdot t + \varphi_0)$ | m |

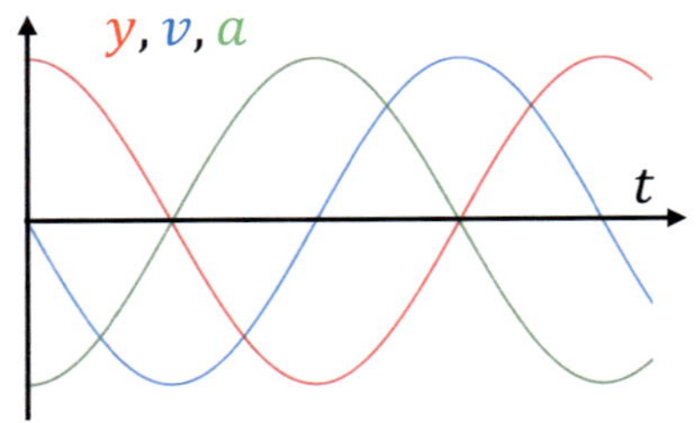


| | | |
|---|---|---|
| Speed | $v = -\omega \cdot \hat{y} \cdot sin(\omega \cdot t + \varphi_0)$ | $\text{m} \cdot \text{s}^{-1}$ |
| Speed amplitude | $\hat{v} = \omega \cdot \hat{y}$ | $\text{m} \cdot \text{s}^{-1}$ |
| Acceleration | $a = -\omega^2 \cdot \hat{y} \cdot cos(\omega \cdot t + \varphi_0)$ | $\text{m} \cdot \text{s}^{-2}$ |
| Acceleration amplitude | $\hat{a} = \omega^2 \cdot \hat{y}$ | $\text{m} \cdot \text{s}^{-2}$ |
| Acceleration-elongation relation | $a = -\omega^2 \cdot y$ | $\text{m} \cdot \text{s}^{-2}$ |

| | | | |
|---|---|---|---|
| Spring pendulum | $\omega = \sqrt{\dfrac{k}{m}}$ <br> $T = 2 \cdot \pi \cdot \sqrt{\dfrac{m}{k}}$ | $k$: spring constant <br> $m$: mass | $\text{s}^{-1}$ <br><br> $\text{s}$ |

| | | | |
|---|---|---|---|
| String pendulum | $\omega = \sqrt{\dfrac{g}{\ell}}$ <br> $T = 2 \cdot \pi \cdot \sqrt{\dfrac{\ell}{g}}$ | 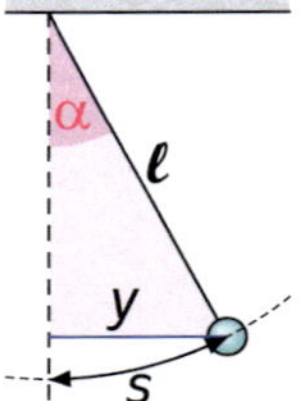 | if $y \ll \ell$ <br> $y$: elongation <br> $g$: acceleration due to gravity <br> $\ell$: length | $\text{s}^{-1}$ <br><br><br> $\text{s}$ |



## Damping

$$y(t) = \hat{y} \cdot exp(-\delta \cdot t) \cdot cos(\omega \cdot t + \varphi_0) \qquad \text{m}$$

| | | |
|---|---|---|
| Lightly damped oscillation | 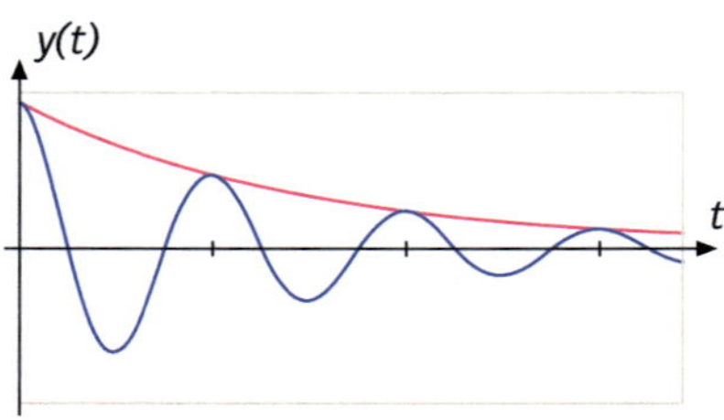 | $\delta$: damping coefficient |



| | | | |
|---|---|---|---|
| Time constant | $\tau = \dfrac{1}{\delta}$ | $\delta$: damping coefficient | s |

# 8. Waves

| | | | |
|---|---|---|---|
| Time | $t$ | | s |
| Amplitude | $\hat{y}$ | | m |
| Period | $T$ | | s |
| Wavelength | $\lambda$ | | m |
| Power of a wave | $P$ | | W |
| Frequency | $f = \dfrac{1}{T}$ | | Hz |
| Angular speed | $\omega = \dfrac{2 \cdot \pi}{T} = 2 \cdot \pi \cdot f$ | | $\text{s}^{-1}$ |
| Intensity of a wave | $I = \dfrac{P}{A}$ | $P$: power of the wave<br>$A$: area pervaded | $\text{W} \cdot \text{m}^{-2}$ |
| Intensity of a point source | $I = \dfrac{P_0}{4 \cdot \pi \cdot r^2}$ | $P_0$: power of the source<br>$r$: distance to the source | $\text{W} \cdot \text{m}^{-2}$ |

## Propagation speed of waves

| | | | |
|---|---|---|---|
| Sound waves in gases | $c = \sqrt{\dfrac{\kappa \cdot R \cdot T}{M_{mol}}}$ | $\kappa$: adiabatic coefficient<br>$R$: universal gas constant<br>$T$: absolute temperature<br>$M_{mol}$: molar mass | $\text{m} \cdot \text{s}^{-1}$ |
| Sound waves in liquids | $c = \dfrac{1}{\sqrt{\chi \rho}}$ | $\chi$: compressibility<br>$\rho$: density | $\text{m} \cdot \text{s}^{-1}$ |
| Sound waves in solids | $c = \sqrt{\dfrac{E}{\rho}}$ | $E$: elasticity module<br>$\rho$: density | $\text{m} \cdot \text{s}^{-1}$ |
| Water waves in shallow water | $c = \sqrt{g \cdot h}$ | $g$: acceleration due to gravity<br>$h$: water depth | $\text{m} \cdot \text{s}^{-1}$ |
| Rope waves | $c = \sqrt{\dfrac{\sigma}{\rho}}$ | $\sigma$: tensile stress<br>$\rho$: density | $\text{m} \cdot \text{s}^{-1}$ |

## Harmonic waves

| | | | |
|---|---|---|---|
| Wave number | $k = \dfrac{2 \cdot \pi}{\lambda}$ | | $\text{m}^{-1}$ |
| Propagation speed | $c = \dfrac{\lambda}{T} = \dfrac{\omega}{k}$ | | $\text{m} \cdot \text{s}^{-1}$ |

| Path difference | $d = \dfrac{\varphi_0}{k}$ | $\varphi_0$: phase<br>$k$: wave number | m |
|---|---|---|---|
| Elongation | $y(x,t) = \hat{y} \cdot \cos(\omega t - kx + \varphi_0)$ <br> $\qquad\;\; = \hat{y} \cdot \cos(\omega t - k(x - d))$ | $k$: wave number<br>$\varphi_0$: phase<br>$d$: path difference | m |

## Interference

| Interference | $y(x,t) = y_1(x,t) + y_2(x,t)$ <br> $\qquad\;\;\; = \hat{y} \cdot \cos(\omega t - kx) + \hat{y} \cdot \cos(\omega t - kx + \varphi)$ | | m |
|---|---|---|---|
| Constructive interference | $\varphi = 2 \cdot n \cdot \pi$ <br> $\quad = 0, \pm 2 \cdot \pi, \pm 4 \cdot \pi, \pm 6 \cdot \pi, \dots$ | $\varphi$: phase difference | rad |
| | $d = n \cdot \lambda = 0, \pm \lambda, \pm 2 \cdot \lambda, \pm 3 \cdot \lambda, \dots$ | $d$: path difference | m |
| Destructive interference | $\varphi = (2 \cdot n + 1) \cdot \pi$ <br> $\quad = \pm \pi, \pm 3 \cdot \pi, \pm 5 \cdot \pi, \dots$ | $\varphi$: phase difference | rad |
| | $d = \left(n + \dfrac{1}{2}\right) \cdot \lambda$ <br> $\quad = \pm \dfrac{1}{2}\lambda, \pm \dfrac{3}{2}\lambda, \pm \dfrac{5}{2}\lambda, \dots$ | $d$: path difference | m |
| Standing wave | $y(x,t) = \hat{y} \cdot \cos(\omega t - kx) + \hat{y} \cdot \cos(\omega t + kx)$ <br> $\qquad\;\;\; = 2 \cdot \hat{y} \cdot \cos(kx) \cdot \cos(\omega t)$ | | m |

## Acoustics

| Hearing threshold | $I_0 = 10^{-12} \dfrac{W}{m^2}$ | | $W \cdot m^{-2}$ |
|---|---|---|---|
| Sound pressure level | $SPL = 10 \cdot \log_{10}\left(\dfrac{I}{I_0}\right)$ | $I$: Intensity<br>$I_0$: hearing threshold | dB |

## Diffraction

| Diffraction angle | $\alpha = \arctan\left(\dfrac{s}{\ell}\right)$ | | ° |
|---|---|---|---|

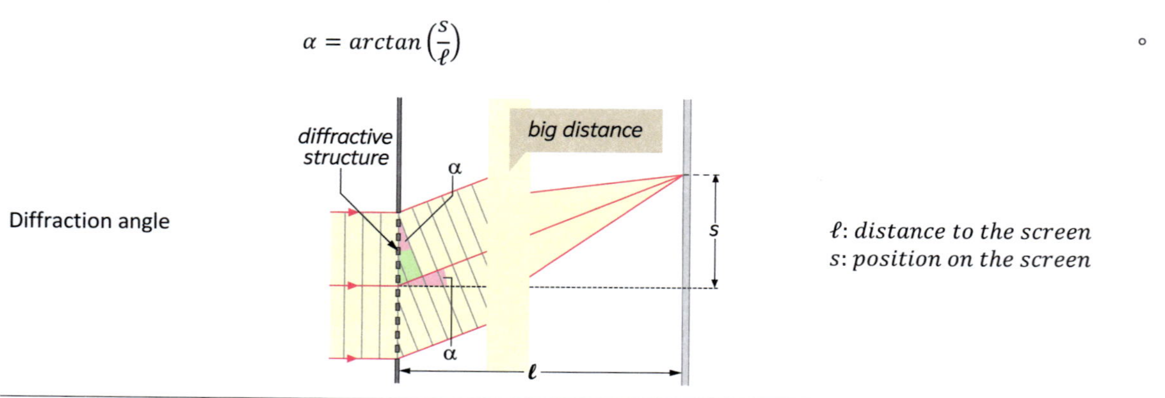


$\ell$: distance to the screen
$s$: position on the screen

$$sin(\alpha) = \frac{\left(n + \frac{1}{2}\right) \cdot \lambda}{b}$$

**Single slit (maxima)**

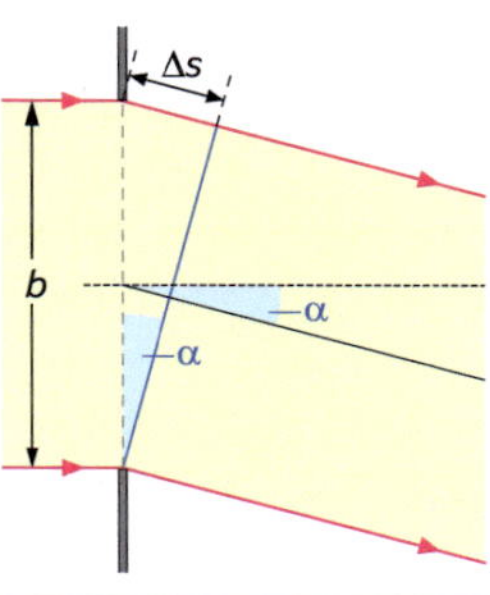


$\alpha$: *diffraction angle*
$n$: *order of the maximum*
$b$: *slit width*

---

**Single slit (minima)**

$$sin(\alpha) = \frac{n \cdot \lambda}{b}$$

$\alpha$: *diffraction angle*
$n$: *order of the minimum*
$b$: *slit width*

---

$$sin(\alpha) = \frac{n \cdot \lambda}{d}$$

**Double slit (maxima)**

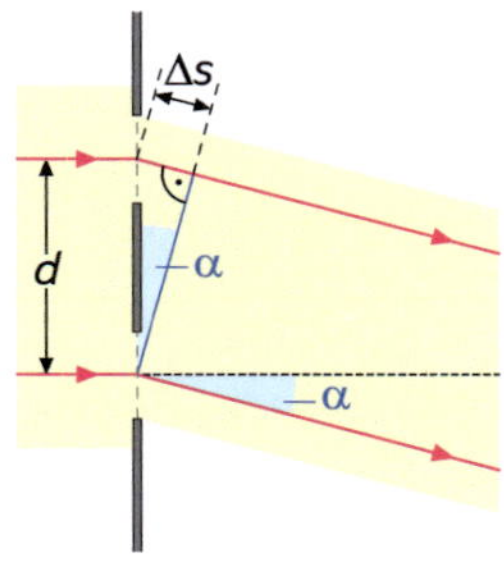


$\alpha$: *diffraction angle*
$n$: *order of the maximum*
$d$: *separation of the slits*

---

**Double slit (minima)**

$$sin(\alpha) = \frac{\left(n + \frac{1}{2}\right) \cdot \lambda}{d}$$

$\alpha$: *diffraction angle*
$n$: *order of the minimum*
$d$: *separation of the slits*

---

$$sin(\alpha) = \frac{n \cdot \lambda}{d}$$

**Grating (main maxima)**

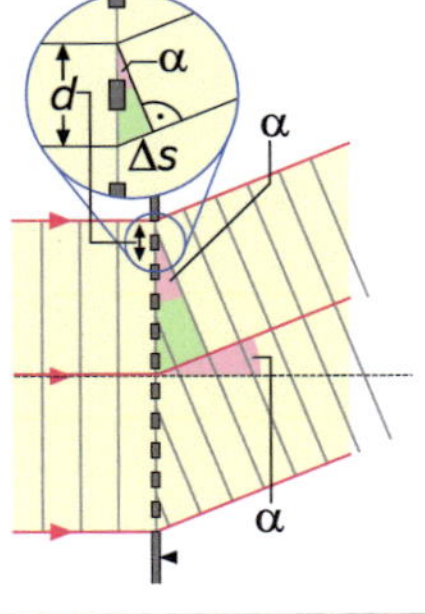


$\alpha$: *diffraction angle*
$n$: *order of the main maximum*
$d$: *separation of the slits*

---

**Pin hole (maxima)**

$$sin(\alpha) = \frac{z \cdot \lambda}{d} \, , where \; z = 0, 1.22, 2.23, 3.24, \dots$$

$\alpha$: *diffraction angle*
$z$: *value for the maximum*
$d$: *diameter of the pin hole*

---

**Pin hole (minima)**

$$sin(\alpha) = \frac{z \cdot \lambda}{d} \, , where \; z = 1.64, 2.68, 3.69, \dots$$

$\alpha$: *diffraction angle*
$z$: *value for the minimum*
$d$: *diameter of the pin hole*

# 9. Electricity and Magnetism

## Electrostatics

| | | | |
|---|---|---|---|
| Permittivity of vacuum | $\varepsilon_0$ | | |
| Relative permittivity | $\varepsilon_r$ | | |
| Coulomb force | $F_C = \dfrac{1}{4 \cdot \pi \cdot \varepsilon_0 \cdot \varepsilon_r} \cdot \dfrac{Q_1 \cdot Q_2}{r^2}$ | $Q_1: charge\ of\ object\ 1$<br>$Q_2: charge\ of\ object\ 2$<br>$r: distance$<br>$\vec{r}: displacement$ | N |
| Coulomb force vector | $\vec{F}_C = \dfrac{1}{4 \cdot \pi \cdot \varepsilon_0 \cdot \varepsilon_r} \cdot \dfrac{Q_1 \cdot Q_2}{r^2} \cdot \dfrac{\vec{r}}{r}$ | $Q_1: charge\ of\ object\ 1$<br>$Q_2: charge\ of\ object\ 2$<br>$r: distance$<br>$\vec{r}: displacement$ | N |
| Voltage | $U = \dfrac{W}{q}$ | $W: work\ performed$<br>$\quad on\ a\ charge$<br>$q: charge$ | V |
| Electric field strength | $E = \dfrac{F_E}{q}$ | 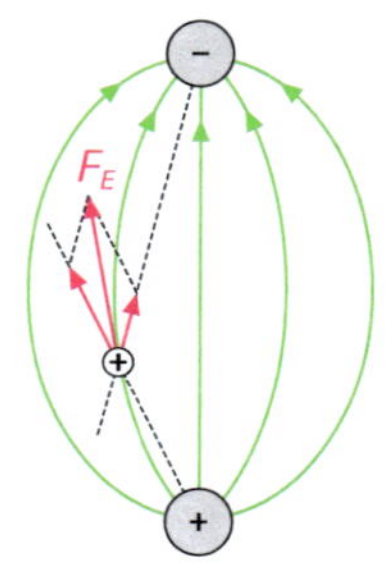<br>$F_E: electric\ force$<br>$\quad on\ a\ charge$<br>$q: charge$ | $N \cdot C^{-1}$ |
| Electric field strength (in a parallel plate capacitor) | $E = \dfrac{U}{d}$ | 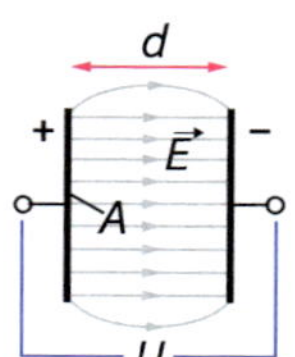<br>$U: voltage\ applied$<br>$d: plate\ distance$ | $N \cdot C^{-1}$ |
| Capacitance (of a parallel plate capacitor) | $C = \dfrac{Q}{U}$ | $Q: charge\ on\ the\ plates$<br>$U: voltage\ applied$ | F |
| Capacitance (of a parallel plate capacitor) | $C = \varepsilon_0 \cdot \varepsilon_r \cdot \dfrac{A}{d}$ | $A: plate\ area$<br>$d: plate\ distance$ | F |
| Energy (of a parallel plate capacitor) | $E_E = \dfrac{1}{2} \cdot C \cdot U^2$ | $C: capacitance$<br>$U: voltage\ applied$ | J |
| Energy density | $\rho_E = \dfrac{1}{2} \cdot \varepsilon_0 \cdot \varepsilon_r \cdot E^2$ | $E: electric\ field\ strength$ | $J \cdot m^{-3}$ |

## Electrodynamics

| Electric current | $I = \dfrac{\Delta Q}{\Delta t}$ | $\dfrac{\Delta Q}{\Delta t}$: *charge flowing per time* | A |
|---|---|---|---|
| Electrical resistance | $R = \dfrac{U}{I}$ | $U$: *voltage applied*<br>$I$: *current flowing* | $\Omega$ |
| Ohm's law | $R = const.$ | | $\Omega$ |
| Voltages in the parallel circuit | $U = U_1 = U_2$ | 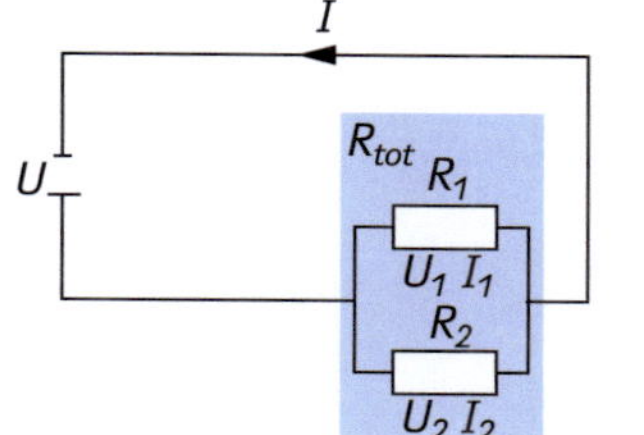 | V |
| Currents in the parallel circuit | $I = I_1 + I_2$ | | A |
| Resistances in the parallel circuit | $R = \left(\dfrac{1}{R_1} + \dfrac{1}{R_2}\right)^{-1}$ | | $\Omega$ |
| Voltages in the series circuit | $U = U_1 + U_2$ | 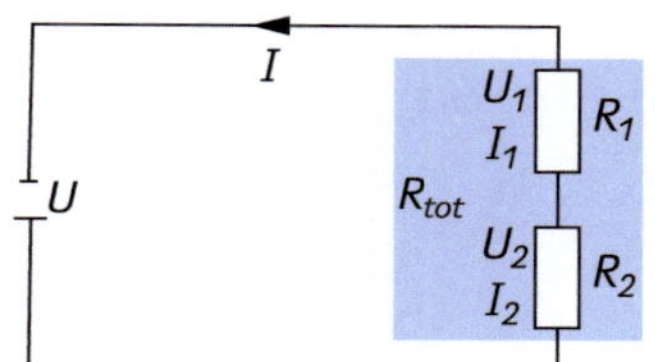 | V |
| Currents in the series circuit | $I = I_1 = I_2$ | | A |
| Resistances in the series circuit | $R = R_1 + R_2$ | | $\Omega$ |
| Electrical work | $W = U \cdot I \cdot t$ | $U$: *voltage applied*<br>$I$: *current flowing*<br>$t$: *time* | J |
| Electric power | $P = U \cdot I$ | $U$: *voltage applied*<br>$I$: *current flowing*<br>$t$: *time* | W |
| Power at a resistor | $P = R \cdot I^2$ | $R$: *resistance of the resistor*<br>$I$: *current flowing* | W |
| Power at a resistor | $P = \dfrac{U^2}{R}$ | $U$: *voltage applied*<br>$R$: *resistance of the resistor* | W |
| Resistivity (specific resistance) | $\rho_{el} = R \cdot \dfrac{A}{\ell}$ | $R$: *resistance*<br>$A$: *cross section*<br>$\ell$: *length* | $\Omega \cdot$ m |

| Kirchhoff's first law (for any node) | $$\sum_{i=1}^{n} I_i = I_1 + I_2 + \cdots + I_n = 0$$ $(\mp$ *signs according to directions*$)$ | 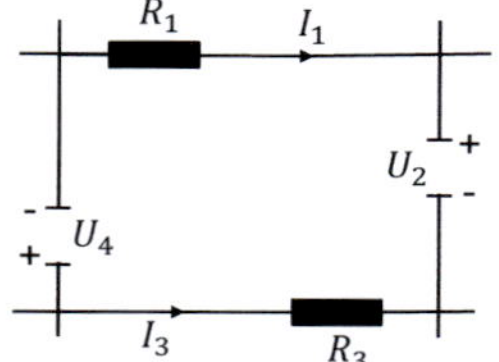 |

| Kirchhoff's second law (for any loop) | $$\sum_{i=1}^{n} U_i = \sum_{i=1}^{m} R_i \cdot I_i$$ $(\mp$ *signs according to directions*$)$ | |

## Magnetism

| Permeability of vacuum | $\mu_0$ | |

| Relative permeability | $\mu_r$ | |

| Magnetic flux density in a coil | $$B = \mu_0 \cdot \mu_r \cdot \frac{n \cdot I}{\ell}$$ 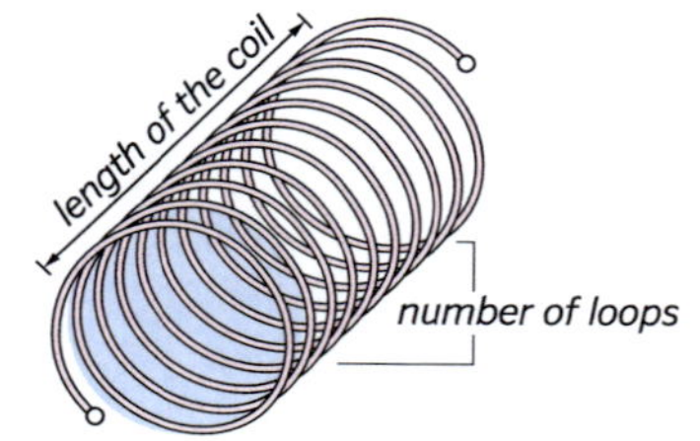 | *n: number of loops* *I: electric current* *$\ell$: length of the coil* | T |



| Magnetic flux density around a conducting wire | $$B = \mu_0 \cdot \mu_r \cdot \frac{I}{2 \cdot \pi \cdot r}$$ 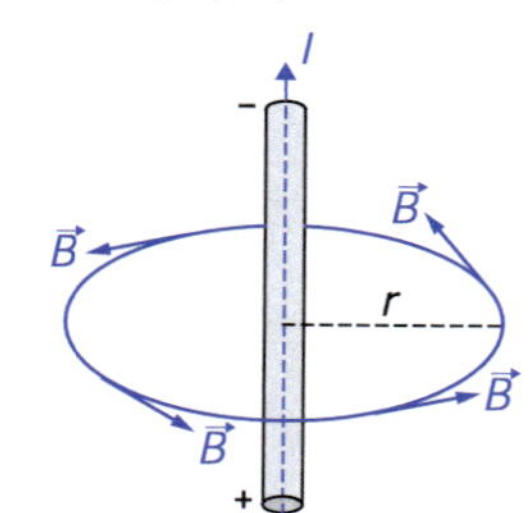 | *I: electric current* *r: distance from the wire* 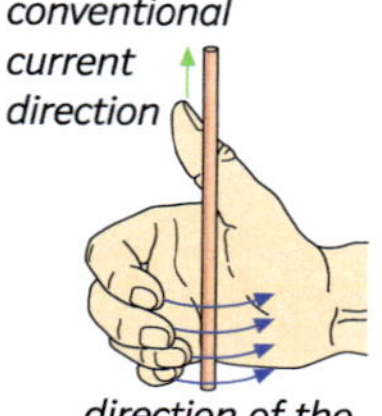 | T |



| Lorentz force | $F_L = B \cdot q \cdot v \qquad$ *if $\vec{B} \perp \vec{v}$* | 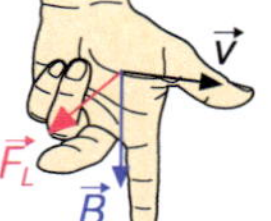 | N |

| Lorentz force | $$\vec{F_L} = q \cdot (\vec{v} \times \vec{B})$$ $$F_L = B \cdot q \cdot v \cdot \sin \alpha$$ | *B: magnetic flux density* *q: charge* *v: electron speed* *$\alpha$: angle between $\vec{v}$ and $\vec{B}$* | N |

| Electromotive force | $\vec{F}_{EMF} = I \cdot (\vec{s} \times \vec{B})$ <br><br> $F_{EMF} = I \cdot s \cdot B \cdot \sin \alpha$ | $B$: magnetic flux density <br> $I$: electric current <br> $s$: wire length in the field B <br> $\alpha$: angle between $\vec{s}$ and $\vec{B}$ | N |
|---|---|---|---|
| Electromotive force | $F_{EMF} = I \cdot s \cdot B \qquad if\ \vec{B} \perp \vec{s}$ | | N |

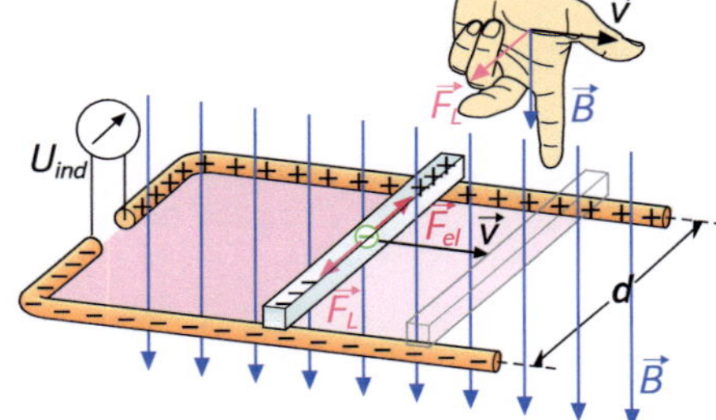


## Induction

| | | | |
|---|---|---|---|
| Induced voltage | $U_{ind} = B \cdot d \cdot v \cdot \sin \alpha$ | $B$: magnetic flux density <br> $d$: wire length <br> $v$: wire speed <br> $\alpha$: angle between $\vec{v}$ and $\vec{B}$ | V |
| Induced voltage | $U_{ind} = B \cdot d \cdot v$ | $if\ \vec{B} \perp \vec{v}$ | V |
| Magnetic flux | $\Phi = B \cdot A$ | $B$: magnetic flux density <br> $A$: area pervaded by B | Wb |
| Induced voltage | $U_{ind} = -\dfrac{\Delta \Phi}{\Delta t}$ | $\Phi$: magnetic flux | V |
| Generator voltage (AC) | $U_{ind}(t) = U_0 \cdot sin(\omega \cdot t)$ | $U_0$: maximum voltage <br> $\omega$: angular speed of the generator <br> $t$: time | V |
| Generator current (AC) | $I(t) = I_0 \cdot sin(\omega \cdot t)$ | $I_0$: maximum current <br> $\omega$: angular speed of the generator <br> $t$: time | A |
| Transformer voltage relation | $\dfrac{U_2}{U_1} = \dfrac{n_2}{n_1}$ | $U_1$: primary voltage <br> $U_2$: secondary voltage <br> $n_1$: primary number of loops <br> $n_2$: secondary number of loops | |

| Transformer power relation | $P_2 = P_1$ | $P_1$: primary power<br>$P_2$: secondary power |
|---|---|---|
| Transformer current relation | $\dfrac{I_2}{I_1} = \dfrac{n_1}{n_2}$ | $I_1$: primary current<br>$I_2$: secondary current<br>$n_1$: primary number of loops<br>$n_2$: secondary number of loops |

# 10. Optics

## Reflection, refraction and shadow

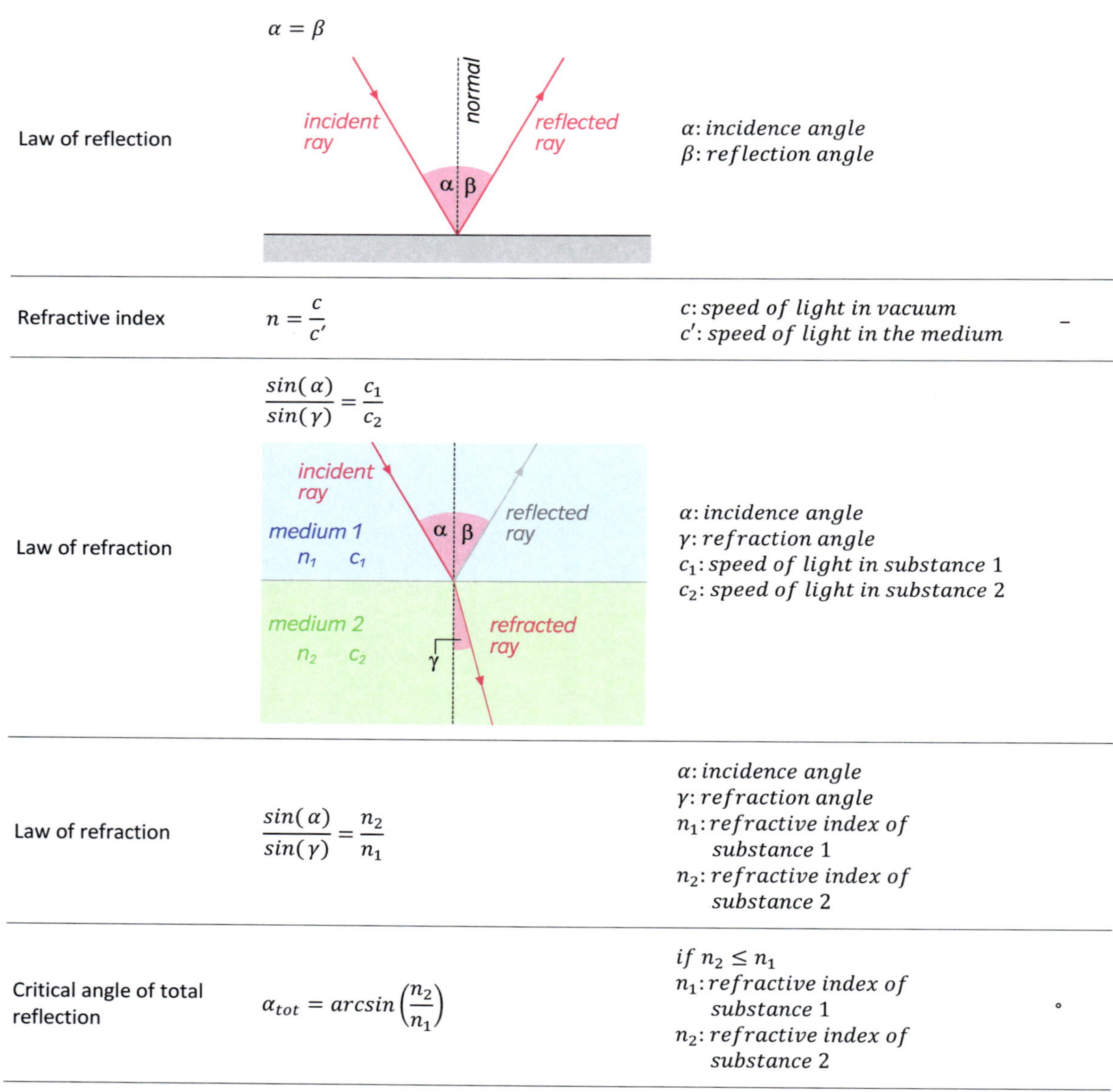


| Law of reflection | $\alpha = \beta$ | $\alpha$: incidence angle<br>$\beta$: reflection angle |
|---|---|---|
| Refractive index | $n = \dfrac{c}{c'}$ | $c$: speed of light in vacuum<br>$c'$: speed of light in the medium |
| Law of refraction | $\dfrac{sin(\alpha)}{sin(\gamma)} = \dfrac{c_1}{c_2}$ | $\alpha$: incidence angle<br>$\gamma$: refraction angle<br>$c_1$: speed of light in substance 1<br>$c_2$: speed of light in substance 2 |
| Law of refraction | $\dfrac{sin(\alpha)}{sin(\gamma)} = \dfrac{n_2}{n_1}$ | $\alpha$: incidence angle<br>$\gamma$: refraction angle<br>$n_1$: refractive index of substance 1<br>$n_2$: refractive index of substance 2 |
| Critical angle of total reflection | $\alpha_{tot} = arcsin\left(\dfrac{n_2}{n_1}\right)$ | if $n_2 \leq n_1$<br>$n_1$: refractive index of substance 1<br>$n_2$: refractive index of substance 2 |

| Shadow formation equation | $\dfrac{h_o}{d_o} = \dfrac{h_i}{d_i}$ | $h_o$: object size<br>$h_i$: image size<br>$d_o$: object distance<br>$d_i$: image distance | |

## Lenses and lens combinations

| Optical power of a lens | $P_{opt} = \dfrac{1}{f}$ | $f$: focal length | dpt |
| Focal length of a lens combination | $f = \left(\dfrac{1}{f_1} + \dfrac{1}{f_2}\right)^{-1}$ | $f$: focal length | m |
| Optical power of a lens combination | $P_{opt} = P_{opt,1} + P_{opt,2}$ | $P_{opt,1}$: optical power of lens 1<br>$P_{opt,2}$: optical power of lens 2 | dpt |

Image formation equation (lens)

$$\frac{h_o}{d_o} = \frac{h_i}{d_i}$$

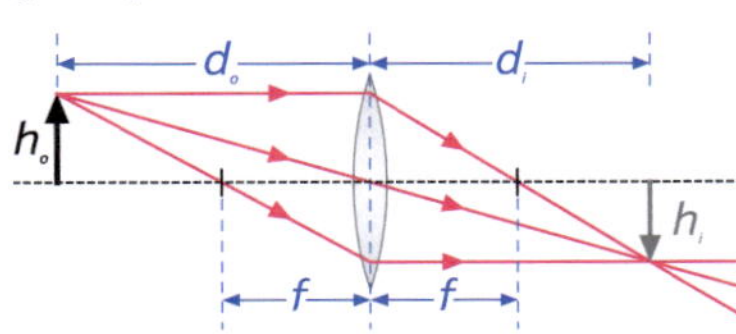

$h_o$: object size
$h_i$: image size
$d_o$: object distance
$d_i$: image distance

| Lens equation | $\dfrac{1}{f} = \dfrac{1}{d_o} + \dfrac{1}{d_i}$ | $f$: focal length<br>$d_o$: object distance<br>$d_i$: image distance | $m^{-1}$ |
| Magnification (lens) | $m = \dfrac{h_i}{h_o}$ | $h_o$: object size<br>$h_i$: image size | – |

## Mirrors

| Focal length of a concave mirror | $f \approx \dfrac{r}{2}$ | $r$: mirror radius | m |

Image formation equation (concave mirror)

$$\frac{h_o}{d_o} = \frac{h_i}{d_i}$$

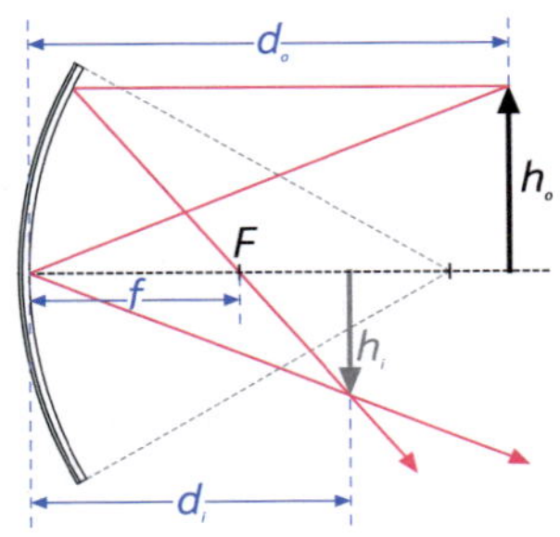

$h_o$: object size
$h_i$: image size
$d_o$: object distance
$d_i$: image distance

| Concave mirror equation | $\dfrac{1}{f} = \dfrac{1}{d_o} + \dfrac{1}{d_i}$ | $f$: focal length<br>$d_o$: object distance<br>$d_i$: image distance | $m^{-1}$ |

| | | | |
|---|---|---|---|
| Magnification (concave mirror) | $m = \dfrac{h_i}{h_o}$ | $h_o$: object size<br>$h_i$: image size | – |
| Focal length of a convex mirror | $f \approx \dfrac{r}{2}$ | $r$: mirror radius | m |

## Pinhole camera

$$\frac{h_o}{d_o} = \frac{h_i}{d_i}$$

| | | | |
|---|---|---|---|
| Image formation equation (pinhole camera) | 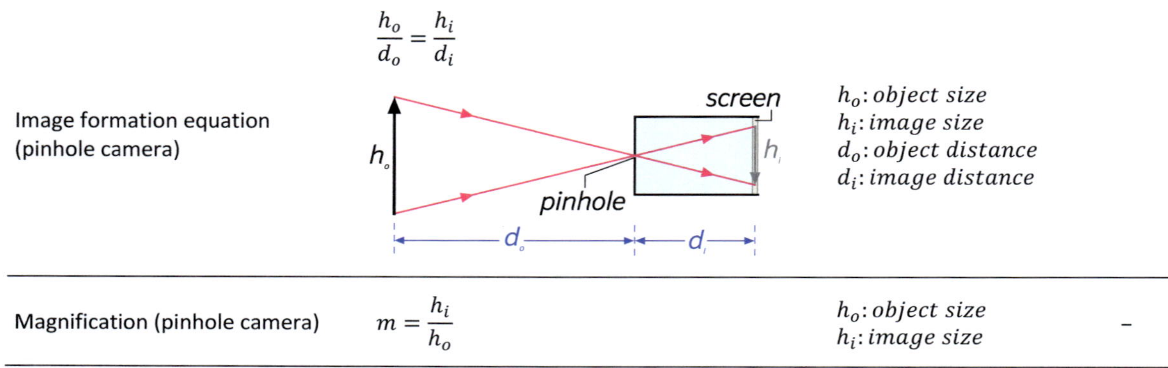 | $h_o$: object size<br>$h_i$: image size<br>$d_o$: object distance<br>$d_i$: image distance | |



| | | | |
|---|---|---|---|
| Magnification (pinhole camera) | $m = \dfrac{h_i}{h_o}$ | $h_o$: object size<br>$h_i$: image size | – |

## Optical instruments

| | | | |
|---|---|---|---|
| Magnification (optical instrument) | $m = \dfrac{\alpha'}{\alpha}$ <br><br>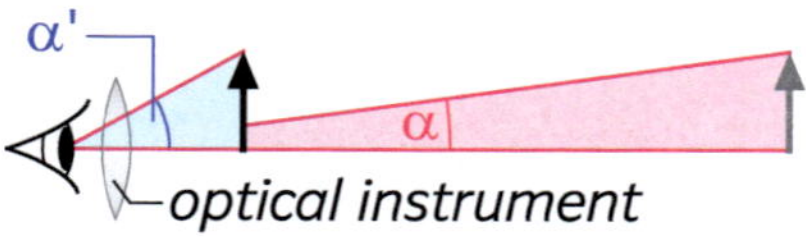 | $\alpha$: field of view<br>$\alpha'$: field of view using the optical instrument | – |



| | | | |
|---|---|---|---|
| Magnification (magnifying glass) | $m = \dfrac{d_o}{f}$ | $f$: focal length<br>$d_o$: object distance from eye; standard: $d_o = 0.25m$ | – |
| Magnification (Kepler telescope) | $m = \dfrac{f_o}{f_e}$ | $f_o$: focal length of the objective<br>$f_e$: focal length of the eyepiece | – |
| Lens distance in the Kepler telescope | $d = f_o + f_e$ | $f_o$: focal length of the objective<br>$f_e$: focal length of the eyepiece | m |
| Magnification (Galileo's telescope) | $m = \dfrac{f_o}{\lvert f_e \rvert}$ | $f_o$: focal length of the objective<br>$f_e$: focal length of the eyepiece | – |
| Lens distance in Galileo's telescope | $d = f_o + f_e = f_o - \lvert f_e \rvert$ | $f_o$: focal length of the objective<br>$f_e$: focal length of the eyepiece | m |
| Magnification (microscope) | $m = m_e \cdot m_o$ | $m_e$: magnification of the eyepiece<br>$m_o$: magnification of the objective | – |

# 11. Radioactivity

| Radiation - absorbed dose | $D = \dfrac{E}{m}$ | $E$: energy absorbed by an object <br> $m$: mass of the object | Gy |
|---|---|---|---|
| Radiation - equivalent dose | $H = \dfrac{E}{m} \cdot \begin{cases} 20 & \alpha - radiation \\ 1 & \beta - radiation \\ 1 & \gamma - radiation \end{cases}$ | $E$: energy absorbed by an object <br> $m$: mass of the object | Sv |

$$N(t) = N_0 \cdot exp(-\lambda \cdot t)$$

| Radioactive decay law | 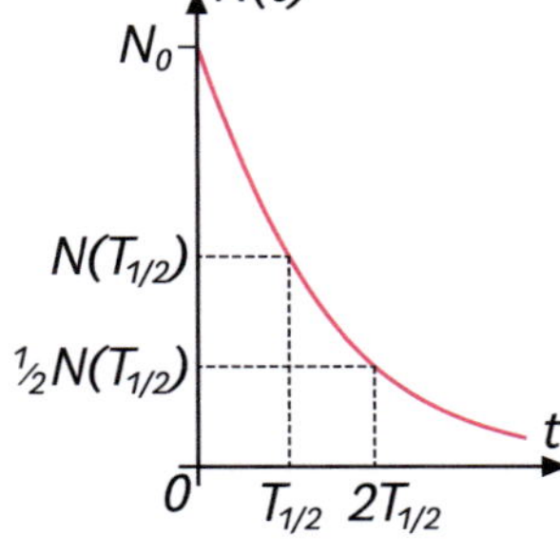 | $N_0$: number of nuclei at $t = 0$ <br> $\lambda$: decay constant <br> $t$: time | – |
|---|---|---|---|



| Decay constant | $\lambda = \dfrac{ln(2)}{T_{1/2}}$ | $T_{1/2}$: half − life | $\text{s}^{-1}$ |
|---|---|---|---|
| Activity | $A = \dfrac{\Delta N}{\Delta t}$ | $\Delta N$: number of decaying nuclei <br> $\Delta t$: time | Bq |
| Initial activity | $A_0 = \lambda \cdot N_0$ | $N_0$: number of nuclei at $t = 0$ <br> $\lambda$: decay constant <br> $t$: time | Bq |
| Activity | $A(t) = A_0 \cdot exp(-\lambda \cdot t)$ | $A_0$: initial activity $(at\ t = 0)$ <br> $\lambda$: decay constant <br> $t$: time | Bq |

# 12. Quantum physics

| Energy of a photon | $E = h \cdot f$ | $h$: Planck constant <br> $f$: frequency of the photon | J |
|---|---|---|---|
| De Broglie relation | $p = \dfrac{h}{\lambda}$ | $h$: Planck constant <br> $\lambda$: wavelength | $\text{N} \cdot \text{s}$ |
| Radiation intensity | $I = \dfrac{N \cdot h \cdot f}{A \cdot t}$ | $N$: number of photons <br> $A$: area pervaded | $\text{W} \cdot \text{m}^{-2}$ |

# 13. Relativity

| | | | |
|---|---|---|---|
| Speed of light | $c$ | | $\mathrm{m \cdot s^{-1}}$ |

Speed of system S' with respect to system S  where S' moves along the x-axis of system S    $v$    $\mathrm{m \cdot s^{-1}}$

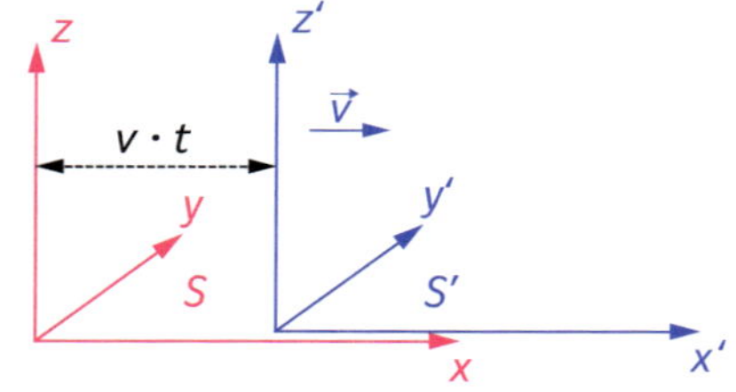


## Classical relativity

### *Galilean transformation*

| | | | |
|---|---|---|---|
| x-position | $x' = x - v \cdot t$ | | m |
| y-position | $y' = y$ | | m |
| z-position | $z' = z$ | | m |
| Time | $t' = t$ | | s |
| Velocity | $\vec{u}' = \vec{u} - \vec{v}$ | $\vec{u}$ : velocity of an object in S<br>$\vec{u}'$ : velocity of the same object in S' | $\mathrm{m \cdot s^{-1}}$ |

### *Inverse Galilean transformation for a system S' moving at speed v with respect to a system S along the x-axis*

| | | | |
|---|---|---|---|
| x-position | $x = x' + v \cdot t'$ | | m |
| y-position | $y = y'$ | | m |
| z-position | $z = z'$ | | m |
| Time | $t = t'$ | | s |
| Velocity | $\vec{u} = \vec{u}' + \vec{v}$ | $\vec{u}$ : velocity of an object in S<br>$\vec{u}'$ : velocity of the same object in S' | $\mathrm{m \cdot s^{-1}}$ |

## Special theory of relativity

| | | | |
|---|---|---|---|
| Beta factor<br>(Speed in the unit c) | $\beta = \dfrac{v}{c}$ | | – |
| Lorentz factor | $\gamma = \dfrac{1}{\sqrt{1 - \beta^2}} = \dfrac{1}{\sqrt{1 - \dfrac{v^2}{c^2}}}$ | $\beta$: beta factor | – |

| | | | |
|---|---|---|---|
| Time dilation | $t' = \dfrac{t}{\gamma}$ | $\gamma$: *Lorentz factor*<br>$t$: *proper time (in S)*<br>$t'$: *time (in S')* | s |
| Length contraction | $\ell = \dfrac{\ell_0}{\gamma}$ | $\gamma$: *Lorentz factor*<br>$\ell_0$: *proper length (in S)*<br>$\ell$: *length (in S')* | m |
| Relativistic mass | $m = \gamma \cdot m_0$ | $\gamma$: *Lorentz factor*<br>$m_0$: *rest mass (in S)*<br>$m$: *relativistic mass (in S')* | kg |
| Mass-energy equivalence | $E_0 = m_0 \cdot c^2$ | $E_0$: *rest energy*<br>$m_0$: *rest mass* | m |
| | $E = m \cdot c^2 = \gamma \cdot m_0 \cdot c^2$ | $E$: *total energy*<br>$m$: *relativistic mass*<br>$\gamma$: *Lorentz factor*<br>$m_0$: *rest mass* | J |
| Relativistic kinetic energy | $E_k = E - E_0 = m \cdot c^2 - m_0 \cdot c^2$ | $E$: *total energy*<br>$E_0$: *rest energy*<br>$m$: *relativistic mass*<br>$m_0$: *rest mass* | J |
| Relativistic momentum | $p = m \cdot v = \gamma \cdot m_0 \cdot v$ | $m$: *relativistic mass*<br>$\gamma$: *Lorentz factor*<br>$m_0$: *rest mass* | N $\cdot$ s |
| Relativistic energy-momentum relation | $E^2 = p^2 \cdot c^2 + m_0{}^2 \cdot c^4$ | $p$: *relativistic momentum*<br>$m_0$: *rest mass* | J |
| Redshift (departing source) | $\lambda_D = \sqrt{\dfrac{1+\beta}{1-\beta}} \cdot \lambda_S$ | $\beta$: *beta factor*<br>$\lambda_S$: *wavelength in the source system*<br>$\lambda_D$: *wavelength in the detector system* | m |
| Blueshift (approaching source) | $\lambda_D = \sqrt{\dfrac{1-\beta}{1+\beta}} \cdot \lambda_S$ | $\beta$: *beta factor*<br>$\lambda_S$: *wavelength in the source system*<br>$\lambda_D$: *wavelength in the detector system* | m |

# III.    Physics Tables

## 1.  Constants

### Fundamental constants

| Physical constant | Symbol | Value |
| --- | --- | --- |
| Atomic mass unit | $u$ | $1.66054 \cdot 10^{-27} \text{kg}$ |
| Avogadro number (exact) | $N_A$ | $6.02214076 \cdot 10^{23} \text{mol}^{-1}$ |
| Boltzmann constant (exact) | $k$ | $1.380649 \cdot 10^{-23} \text{J} \cdot \text{K}^{-1}$ |
| Charge-to-mass ratio for the electron | $e/m_e$ | $1.75882 \cdot 10^{11} \text{C} \cdot \text{kg}^{-1}$ |
| Elementary charge (exact) | $e$ | $1.602176634 \cdot 10^{-19} \text{C}$ |
| Gas constant (exact: $R = k \cdot N_A$) | $R$ | $8.314462618 \, \text{N} \cdot \text{m} \cdot \text{K}^{-1} \cdot \text{mol}^{-1}$ |
| Gravitational constant | $G$ | $6.67430 \cdot 10^{-11} \text{N} \cdot \text{m}^2 \cdot \text{kg}^{-2}$ |
| Mass of the electron | $m_e$ | $9.1093897 \cdot 10^{-31} \text{kg}$ |
| Mass of the neutron | $m_n$ | $1.6749286 \cdot 10^{-27} \text{kg}$ |
| Mass of the proton | $m_p$ | $1.6726231 \cdot 10^{-27} \text{kg}$ |
| Molar volume * | $V_{mol}$ | $2.24138 \cdot 10^{-2} \text{m}^3$ |
| Permeability of vacuum | $\mu_0$ | $1.2566370614 \cdot 10^{-6} \text{V} \cdot \text{s} \cdot \text{A}^{-1} \cdot \text{m}^{-1}$ |
| Permittivity of vacuum | $\varepsilon_0$ | $8.854187817 \cdot 10^{-12} \text{A} \cdot \text{s} \cdot \text{V}^{-1} \cdot \text{m}^{-1}$ |
| Planck constant (exact) | $h$ | $6.62607015 \cdot 10^{-34} \text{J} \cdot \text{s}$ |
| Speed of light in vacuum (exact) | $c$ | $2.99792458 \cdot 10^{8} \text{m} \cdot \text{s}^{-1}$ |
| Stefan-Boltzmann constant (exact $\sigma = 2 \cdot \pi^5 \cdot k^4 \cdot (15 \cdot h^3 \cdot c^2)^{-1}$) | $\sigma$ | $5.670374419 \cdot 10^{-8} \text{W} \cdot \text{m}^{-2} \cdot \text{K}^{-4}$ |

* at standard temperature

### Standard conditions

| | | | |
| --- | --- | --- | --- |
| Standard temperature | $T_S, \vartheta_S$ | $273.15 \text{K}$ | $0\,°C$ |
| Standard pressure | $p_S$ | $101325 \text{Pa}$ | $1.01325 \text{bar}$ |
| Standard acceleration due to gravity | $g_S$ | $9.80665 \text{m} \cdot \text{s}^{-2}$ | – |

## 2.  Units

### SI-units

#### *SI-base units (current definitions)*

A **second** is the time of 9192631770 periods of the radiation corresponding to the transition between the two hyperfine levels of the ground state of the Cs-133 atom at 0 K.

A **metre** is the distance that light travels in 1/299792458 of a second in vacuum.

The **kilogram** is defined by taking the Planck constant h to be $6.62607015 \cdot 10^{-34} \mathrm{kg} \cdot \mathrm{m}^2 \cdot \mathrm{s}^{-1}$.

The **kelvin** is defined by taking the Boltzmann constant k to be $1.380649 \cdot 10^{-23} \mathrm{kg} \cdot \mathrm{m}^2 \cdot \mathrm{s}^{-2} \cdot \mathrm{K}^{-1}$.

The **mole** is defined by taking the Avogadro constant $N_A$ to be $6.02214076 \cdot 10^{23} \mathrm{mol}^{-1}$.

The **ampere** is defined by taking the elementary charge e to be $1.602176634 \cdot 10^{-19} \mathrm{A} \cdot \mathrm{s}$.

A **candela** is the luminous intensity, in a given direction, of a source that emits monochromatic radiation of frequency 540 THz and that has a radiant intensity in that direction of 1/683 watt per steradian.

#### *SI-base units (older definitions)*

A **second** is the time of 9192631770 periods of the radiation corresponding to the transition between the two hyperfine levels of the ground state of the Cs-133 atom at 0 K.

A **metre** is the length of the prototype metre bar.

The **kilogram** is the mass of the prototype kilogram

A **kelvin** is 1/273.15 of the thermodynamic temperature of the triple point of water.

A **mole** is the amount of substance of a system that contains as many "elementary entities" (e.g. atoms, molecules, ions, electrons) as there are atoms in 12 g of C-12.

An **ampere** is the constant current needed to produce a force of $2 \cdot 10^{-7} \mathrm{N}$ per metre between two straight parallel conductors of infinite length and negligible circular cross-section placed one metre apart in vacuum.

A **candela** is the luminous intensity, in a given direction, of a source that emits monochromatic radiation of frequency 540 THz and that has a radiant intensity in that direction of 1/683 watt per steradian.

### *SI-base units*

| Base quantity | Variable | Base unit | Symbol |
|---|---|---|---|
| Length | $s, \ell, r$ | metre | m |
| Time | $t$ | second | s |
| Mass | $m$ | kilogram | kg |
| Temperature | $T$ | kelvin | K |
| Amount of substance | $n$ | mole | mol |
| Electric current | $I$ | ampere | A |
| Luminous intensity | $I_L$ | candela | cd |

## Prefixes

| Power of Ten | Prefix | Abbreviation |
|---|---|---|
| $10^{18}$ | exa | E |
| $10^{15}$ | peta | P |
| $10^{12}$ | tera | T |
| $10^{9}$ | giga | G |
| $10^{6}$ | mega | M |
| $10^{3}$ | kilo | k |
| $10^{0}$ | - | - |
| $10^{-3}$ | milli | m |
| $10^{-6}$ | micro | μ |
| $10^{-9}$ | nano | n |
| $10^{-12}$ | pico | p |
| $10^{-15}$ | femto | f |
| $10^{-18}$ | atto | a |

| Power of Ten | Prefix | Abbreviation |
|---|---|---|
| $10^{2}$ | hecto | h |
| $10^{1}$ | deka | da |
| $10^{0}$ | - | - |
| $10^{-1}$ | deci | d |
| $10^{-2}$ | centi | c |

## Derived SI-units

| Physical quantity | Name of the unit | Variable | Symbol | Relation to SI-base-units |
|---|---|---|---|---|
| Area | square metre | $A$ | $\mathrm{m^2}$ | $\mathrm{m^2}$ |
| Volume | cubic metre | $V$ | $\mathrm{m^3}$ | $\mathrm{m^3}$ |
| Density | kilogram per cubic metre | $\rho$ | $\mathrm{kg \cdot m^{-3}}$ | $\mathrm{kg \cdot m^{-3}}$ |
| Frequency | hertz | $f$ | $\mathrm{Hz}$ | $\mathrm{Hz = s^{-1}}$ |
| Speed | metre per second | $v$ | $\mathrm{m \cdot s^{-1}}$ | $\mathrm{m \cdot s^{-1}}$ |
| Flow | cubic metre per second | $Q$ | $\mathrm{m^3 \cdot s^{-1}}$ | $\mathrm{m^3 \cdot s^{-1}}$ |
| Acceleration | metre per second squared | $a$ | $\mathrm{m \cdot s^{-2}}$ | $\mathrm{m \cdot s^{-2}}$ |
| Force | newton | $F$ | $\mathrm{N}$ | $\mathrm{N = kg \cdot m \cdot s^{-2}}$ |
| Pressure | pascal | $p$ | $\mathrm{Pa}$ | $\mathrm{Pa = N \cdot m^{-2} = kg \cdot m^{-1} \cdot s^{-2}}$ |
| Torque | newton metre | $M$ | $\mathrm{N \cdot m}$ | $\mathrm{N \cdot m = kg \cdot m^2 \cdot s^{-2}}$ |
| Energy | joule | $E$ | $\mathrm{J}$ | $\mathrm{J = N \cdot m = kg \cdot m^2 \cdot s^{-2}}$ |
| Power | watt | $P$ | $\mathrm{W}$ | $\mathrm{W = J \cdot s^{-1} = kg \cdot m^2 \cdot s^{-3}}$ |
| Viscosity | pascal second | $\eta$ | $\mathrm{Pa \cdot s}$ | $\mathrm{Pa \cdot s = kg \cdot m^{-1} \cdot s^{-1}}$ |
| Electric charge | coulomb | $Q$ | $\mathrm{C}$ | $\mathrm{C = A \cdot s}$ |
| Voltage | volt | $U$ | $\mathrm{V}$ | $\mathrm{V = J \cdot C^{-1} = kg \cdot m^2 \cdot s^{-3} \cdot A^{-1}}$ |
| Resistance | ohm | $R$ | $\Omega$ | $\mathrm{\Omega = V \cdot A^{-1} = kg \cdot m^2 \cdot s^{-3} \cdot A^{-2}}$ |
| Capacitance | farad | $C$ | $\mathrm{F}$ | $\mathrm{F = C \cdot V^{-1} = A^2 \cdot s^4 \cdot kg^{-1} \cdot m^{-2}}$ |
| Magnetic flux density; magnetic field strength | tesla | $B$ | $\mathrm{T}$ | $\mathrm{T = N \cdot A^{-1} \cdot m^{-1} = kg \cdot A^{-1} \cdot s^{-2}}$ |
| Magnetic flux | weber | $\Phi$ | $\mathrm{Wb}$ | $\mathrm{Wb = T \cdot m^2 = kg \cdot m^2 \cdot A^{-1} \cdot s^{-2}}$ |
| Inductance | henry | $L$ | $\mathrm{H}$ | $\mathrm{H = V \cdot s \cdot A^{-1} = kg \cdot m^2 \cdot s^{-2} \cdot A^{-2}}$ |
| Radioactivity; Activity | becquerel | $A$ | $\mathrm{Bq}$ | $\mathrm{Bq = s^{-1}}$ |
| Radiation - absorbed dose | gray | $D$ | $\mathrm{Gy}$ | $\mathrm{Gy = J \cdot kg^{-1} = m^2 \cdot s^{-2}}$ |
| Radiation - equivalent dose | sievert | $H$ | $\mathrm{Sv}$ | $\mathrm{Sv = J \cdot kg^{-1} = m^2 \cdot s^{-2}}$ |

## Non-SI-units

| Physical quantity | Name of the unit | Symbol | Definition | Relation to SI-units |
|---|---|---|---|---|
| **Length** | angstrom | Å | $\equiv 10^{-10}$m | $= 10^{-10}$m |
| | astronomical unit | AU | $\approx$ distance from Earth to Sun | $\approx 1.4959787 \cdot 10^{11}$m |
| | foot | ft | $\equiv \frac{1}{3}$yd $= 12$ in | $= 0.3048$m |
| | inch | in | $\equiv \frac{1}{36}$yd $= \frac{1}{12}$ft | $= 0.0254$m |
| | light-year | ly | $\equiv c \cdot 1y = c \cdot 365.25$d | $= 9.4607304725808 \cdot 10^{15}$m |
| | mile | mi | $\equiv 1760$yd $= 5280$ft | $= 1609.344$m |
| | nautical mile | nmi | $\equiv 1852$m | $= 1852$m |
| | yard | yd | $\equiv 0.9144$m $\equiv 3$ft $\equiv 36$in | $= 0.9144$ m |
| **Area** | acre | ac | $\equiv 4840$yd$^2$ | $= 4046.8564224$m$^2$ |
| | are | a | $\equiv 100$m$^2$ | $= 100$m$^2$ |
| | hectare | ha | $\equiv 10000$m$^2$ | $= 10000$m$^2$ |
| **Volume** | barrel (US) | bl | $\equiv 42$gal | $= 0.158987294928$m$^3$ |
| | gallon (US fluid) | gal | $\equiv 231$in$^3$ | $= 3.785411784 \cdot 10^{-3}$m$^3$ |
| | litre | $\ell$ | $\equiv 1$dm$^3$ | $= 10^{-3}$m$^3$ |
| | pint (US fluid) | pt | $\equiv \frac{1}{8}$gal | $= 473.176473 \cdot 10^{-6}$m$^3$ |
| | quart (US fluid) | qt | $\equiv \frac{1}{4}$gal | $= 946.352946 \cdot 10^{-6}$m$^3$ |
| | tablespoon (metric) | tbsp | $\equiv 15$m$\ell$ | $= 1.5 \cdot 10^{-5}$ m$^3$ |
| | teaspoon (metric) | tsp | $\equiv 5$m$\ell$ | $= 5 \cdot 10^{-6}$ m$^3$ |
| **Mass** | carat | kt | $\equiv 3\frac{1}{6}$gr | $\approx 205.196548333$mg |
| | carat (metric) | ct | $\equiv 200$mg | $= 200$mg |
| | grain | gr | $\equiv 64.79891$mg | $= 64.79891$mg |
| | ounce (troy) | oz t | $\equiv \frac{1}{12}$lb t | $= 31.1034768$g |
| | ounce | oz | $\equiv \frac{1}{16}$lb | $= 28.349523125$g |
| | pound | lb | $\equiv 7000$gr | $= 0.45359237$kg |
| | pound (troy) | lb t | $\equiv 5760$gr | $= 0.3732417216$kg |
| | tonne (metric) | t | $\equiv 1000$kg | $= 1000$kg |

## Non-SI-units (continued)

| Physical quantity | Name of the unit | Symbol | Definition | Relation to SI-units |
|---|---|---|---|---|
| **Time** | day | d | $\equiv 24\text{h}$ | $= 86400\text{s}$ |
| | hour | h | $\equiv 60\text{min}$ | $= 3600\text{s}$ |
| | minute | min | $\equiv 60\text{sec}$ | $= 60\text{s}$ |
| | year (Julian) | a or y | $\equiv 365.25\text{d}$ | $= 31557600\text{s}$ |
| **Speed** | kilometre per hour | km/h | $\equiv 1\,\text{km} \cdot \text{h}^{-1}$ | $= \dfrac{1}{3.6}\text{m} \cdot \text{s}^{-1}$ |
| | knot | kn | $\equiv 1\text{nmi} \cdot \text{h}^{-1} = 1.852\text{km} \cdot \text{h}^{-1}$ | $\approx 0.514444\text{m} \cdot \text{s}^{-1}$ |
| | speed of light in vacuum | c | $\equiv 299792458\text{m} \cdot \text{s}^{-1}$ | $= 299792458\text{m} \cdot \text{s}^{-1}$ |
| **Force** | pound-force | lbf | $\equiv \text{lb} \cdot g_S$ | $\approx 4.448221615\text{N}$ |
| **Pressure** | atmosphere (standard) | atm | $\equiv p_S$ | $= 101\,325\text{Pa}$ |
| | bar | bar | $\equiv 10^5\text{Pa}$ | $= 10^5\text{Pa}$ |
| | millimetre of mercury | mmHg | $\equiv 13595.1\,\text{kg} \cdot \text{m}^{-3} \cdot 1\,\text{mm} \cdot g_S$ | $\approx 1.33322 \cdot 10^2\text{Pa}$ |
| | pound-force per square inch | psi | $\equiv \text{lbf} \cdot \text{in}^{-2}$ | $\approx 6894.757293\text{Pa}$ |
| **Energy** | calorie | cal | $\equiv 4.1868\text{J}$ | $= 4.1868\text{J}$ |
| | electron volt | eV | $\equiv 1e \cdot 1\text{V}$ | $\approx 1.60217733 \cdot 10^{-19}\text{J}$ |
| | kilowatt-hour | kWh | $\equiv 1\text{kW} \cdot 1\text{h}$ | $= 3.6 \cdot 10^6\text{J}$ |
| **Power** | horsepower (metric) | hp | $\equiv 75\,\text{kg} \cdot \text{m} \cdot \text{s}^{-1} \cdot g_S$ | $= 735.49875\text{W}$ |
| **Electric charge** | atomic unit of charge | au | $\equiv 1e$ | $\approx 1.60217733 \cdot 10^{-19}\text{C}$ |
| **Temperature** | degree Celsius | °C | $[°C] \equiv [K] - 273.15$ | $[K] \equiv [°C] + 273.15$ |
| | degree Fahrenheit | °F | $[°F] \equiv \dfrac{9}{5} \cdot [K] - 459.65$ | $[K] \equiv ([°F] + 459.67) \cdot \dfrac{5}{9}$ |
| **Radioactivity (Activity)** | curie | Ci | $3.7 \cdot 10^{10}\text{Bq}$ | $= 3.7 \cdot 10^{10}\text{Bq}$ |

# 3.  Mechanics

## Densities

### *Densities of solids at 20°C*

$\rho$: Density

| Solid | $\rho\ [10^3\text{kg} \cdot \text{m}^{-3}]$ | | Solid | $\rho\ [10^3\text{kg} \cdot \text{m}^{-3}]$ |
|---|---|---|---|---|
| Aluminium | 2.71 | | Mica * | 2.9 |
| Amber | 1.08 | | Nickel | 8.90 |
| Beech wood (dry) * | 0.69 | | Oak Wood (dry) * | 0.67 |
| Beryllium | 1.84 | | Opal * | 2.2 |
| Birch wood (dry) * | 0.65 | | Paraffin * | 0.9 |
| Brass * | 8.5 | | Platinum | 21.5 |
| Carbon Steel | 7.83 | | Plexiglass | 1.18 |
| Cast iron * | 7.3 | | Plutonium | 19.8 |
| Clay brick * | 1.6 | | Porcelain * | 2.4 |
| Concrete * | 2.2 | | Quartz * | 2.7 |
| Copper | 8.92 | | Quartz glass * | 2.2 |
| Cork * | 0.25 | | Rubber * | 1.15 |
| Diamond | 3.50 | | Silicon | 2.42 |
| Flint * | 2.6 | | Silver | 10.5 |
| Germanium | 5.35 | | Sodium | 0.971 |
| Gold | 19.3 | | Spruce wood (dry) * | 0.51 |
| Graphite | 2.24 | | Steel | 7.85 |
| Ice (at 0°C) | 0.917 | | Styrofoam * | 0.020 |
| Invar | 8.01 | | Tin | 7.28 |
| Iron | 7.85 | | Titanium | 4.50 |
| Lead | 11.3 | | Tungsten | 19.6 |
| Leather (dry) * | 1.35 | | Uranium | 18.9 |
| Limestone * | 2.7 | | Window Glass * | 2.6 |
| Marble * | 2.7 | | Zinc | 7.13 |

* typical value

## *Densities of liquids at 20°C*

$\rho$: Density

| Liquid | $\rho \left[10^3 \mathrm{kg} \cdot \mathrm{m}^{-3}\right]$ |
| --- | --- |
| Acetone | 0.791 |
| Benzene | 0.878 |
| Bromine | 3.13 |
| Carbon tetrachloride | 1.59 |
| Crude oil * | 0.95 |
| Diethyl ether | 0.714 |
| Ethanol | 0.785 |
| Freon-11 | 1.49 |
| Gas oil * | 0.89 |
| Gasoline, natural * | 0.71 |
| Gasoline, vehicle * | 0.74 |
| Glycerine (Glycerol) | 1.26 |
| Heavy water | 1.11 |
| Linseed oil * | 0.929 |
| Mercury | 13.5 |
| Methanol | 0.790 |
| Milk * | 1.04 |
| Olive oil * | 0.86 |
| Petroleum * | 0.85 |
| Rape seed oil * | 0.92 |
| Sea water * | 1.02 |
| Silicone oil * | 0.76 |
| Soya bean oil * | 0.91 |
| Toluol | 0.867 |
| Water | 0.9983 |

* typical value

## *Densities of gases in standard conditions*

$\rho$: Density

| Gas | $\rho \left[\mathrm{kg} \cdot \mathrm{m}^{-3}\right]$ |
| --- | --- |
| Acetylene | 1.17 |
| Air | 1.293 |
| Ammonia | 0.769 |
| Argon | 1.78 |
| Butane | 2.50 |
| Carbon dioxide | 1.98 |
| Carbon monoxide | 1.25 |
| Chlorine | 3.21 |
| Fluorine | 1.70 |
| Helium | 0.179 |
| Hydrogen | 0.0899 |
| Krypton | 3.75 |
| Methane | 0.717 |
| Natural gas * | 0.8 |
| Neon | 0.90 |
| Nitrogen | 1.25 |
| Oxygen | 1.43 |
| Radon | 9.73 |
| Sulfur dioxide | 2.93 |
| Xenon | 5.90 |

* typical value

# Friction coefficients

## Static and dynamic friction coefficients (typical values)

$\mu_S$: Static friction coefficient

$\mu_D$: Dynamic friction coefficient

| Material Combination | | $\mu_S$ | $\mu_D$ |
|---|---|---|---|
| Steel | Steel | 0.2 | 0.15 |
| Steel | Steel (lubricated) | 0.15 | 0.1 |
| Steel | Wood | 0.4 | 0.2 |
| Steel | Teflon | 0.05 | 0.05 |
| Wood | Wood | 0.6 | 0.4 |
| Wood | Stone | 0.7 | 0.3 |
| Glass | Glass | 0.9 | 0.4 |
| Teflon | Teflon | 0.05 | 0.05 |
| Car tyre | Asphalt (dry) | 0.8 | 0.6 |
| Car tyre | Asphalt (wet, clean) | 0.6 | 0.4 |
| Car tyre | Asphalt (wet, greasy) | 0.3 | 0.2 |
| Car tyre | Asphalt (icy) | <0.2 | <0.15 |
| Ski (waxed) | Snow | 0.1 | 0.05 |
| Ice skate | Ice | 0.03 | 0.01 |

## Rolling friction coefficients (typical values)

$\mu_R$: Rolling friction coefficient (Warning: All values strongly dependent on wheel radius and speed!)

| Material Combination | $\mu_R$ |
|---|---|
| Roundwoods, rollers and wheels on earth road | 0.5 |
| Roundwoods on rough substratum | 0.2 |
| Roundwoods on asphalt, steel or on similar solid substrata | 0.1 |
| Pneumatic-tyred wheels on solid substrata | 0.02 |
| Steel wheels on rails | 0.01 |
| High tech steel wheels on straight clean tracks | 0.0025 |

# Viscosities

## *Viscosity of liquids at 20°C*

$\eta$: Viscosity

| Substance | $\eta\ [10^{-3}\mathrm{Pa} \cdot \mathrm{s}]$ |
|---|---|
| Benzene | 0.601 |
| Cream * | 10 |
| Diethyl ether | 0.240 |
| Ethanol | 1.19 |
| Glycerine (Glycerol) | 1480 |
| Heptane | 0.410 |
| Hexane | 0.320 |
| Honey * | 10000 |
| Mercury | 1.55 |
| Nonane | 0.711 |
| Octane | 0.538 |
| Olive oil * | 100 |
| Paraffin oil * | 100 |
| Pentane | 0.224 |
| Petroleum * | 0.65 |
| Syrup * | 100000 |

* typical value

## *Viscosity of gases at 0°C*

$\eta$: Viscosity

| Substance | $\eta\ [10^{-6}\mathrm{Pa} \cdot \mathrm{s}]$ |
|---|---|
| Air | 17.1 |
| Oxygen | 19.2 |
| Carbon | 13.8 |
| Nitrogen | 16.6 |
| Argon | 21.0 |
| Neon | 29.7 |
| Helium | 18.6 |
| Hydrogen | 8.40 |

## *Viscosity of water at various temperatures*

$\vartheta$: Temperature
$\eta$: Viscosity

| $\vartheta[°\mathrm{C}]$ | $\eta\ [10^{-3}\mathrm{Pa} \cdot \mathrm{s}]$ |
|---|---|
| 10 | 1.308 |
| 20 | 1.002 |
| 30 | 0.7978 |
| 40 | 0.6531 |
| 50 | 0.5471 |
| 60 | 0.4668 |
| 70 | 0.4044 |
| 80 | 0.3550 |
| 90 | 0.3150 |
| 100 | 0.2822 |

## Various mechanical data

### *Drag coefficients*

$E$: Modulus of elasticity

| Object shape | $c_D$ |
|---|---|
| Disc | 1.11 |
| Sphere | 0.4 |
| Half-sphere (concave) | 1.33 |
| Half-sphere (convex) | 0.34 |
| Cone (30°) | 0.34 |
| Cone (60°) | 0.51 |
| Cylinder (long) | 0.85 |
| Streamlined body (long) | 0.1 |
| Compact car * | 0.4 |
| Automobile * | 0.35 |
| Sports car * | 0.28 |
| Parachute * | 0.9 |

* typical value

### *Compressibility of liquids at 20°C*

$\chi$: Compressibility

| Substance | $\chi$ [$10^{-11}$Pa$^{-1}$] |
|---|---|
| Ethanol | 110 |
| Glycerine (Glycerol) | 21 |
| Mercury | 3.7 |
| Water | 45.8 |
| Diethyl ether | 19 |
| Acetone | 120 |
| Oil * | 60 |

* typical value

### *Modulus of elasticity at 20°C*

$c_D$: Drag coefficient

| Material | $E$ [$10^9$N · m$^{-2}$] |
|---|---|
| Aluminium | 71.5 |
| Beryllium | 287 |
| Bone * | 20 |
| Brass * | 110 |
| Concrete * | 30 |
| Copper | 117 |
| Diamond | 1230 |
| Fibreglass * | 70 |
| Glass * | 70 |
| Lead | 17.0 |
| Magnesium | 45.0 |
| Molybdenum | 329 |
| Nickel | 200 |
| Nylon * | 3 |
| Osmium | 550 |
| Silicon | 185 |
| Steel * | 200 |
| Titanium | 105 |
| Tungsten | 411 |
| Wood (against grain) * | 0.71 |
| Wood (along grain) * | 14 |

* typical value

## *Speed of sound at 20°C*

$c$: Speed of sound

| Material | $c\,[\mathrm{m \cdot s^{-1}}]$ |
|---|---|

**Solids (longitudinal waves in thin rods)**

| Material | $c\,[\mathrm{m \cdot s^{-1}}]$ |
|---|---|
| Aluminium | 5160 |
| Brass | 3460 |
| Clay brick | 3700 |
| Concrete | 4100 |
| Glass | 5100 |
| Lead | 1250 |
| Steel | 5110 |
| Wood | 3500 |

**Gases (at standard pressure)**

| Material | $c\,[\mathrm{m \cdot s^{-1}}]$ |
|---|---|
| Air | 344 |
| Carbon dioxide | 267 |
| Helium | 991 |
| Hydrogen | 1290 |
| Methane | 450 |
| Nitrogen | 335 |
| Oxygen | 319 |

**Liquids**

| Material | $c\,[\mathrm{m \cdot s^{-1}}]$ |
|---|---|
| Acetone | 1195 |
| Benzene | 1325 |
| Ethanol | 1160 |
| Glycerine (Glycerol) | 1910 |
| Mercury | 1390 |
| Water | 1475 |

## 4. Astronomy

## The solar system

### The planets

| | Mercury | Venus | Earth | Mars | Jupiter | Saturn | Uranus | Neptune |
|---|---|---|---|---|---|---|---|---|
| Mass [$10^{24}$kg] | 0.3301 | 4.869 | 5.974 | 0.6419 | 1899 | 568.5 | 86.83 | 102.4 |
| Volumetric mean radius [$10^6$m] | 2.441 | 6.052 | 6.371 | 3.390 | 69.91 | 58.23 | 25.36 | 24.62 |
| Equatorial radius [$10^6$m] | 2.440 | 6.052 | 6.3781 | 3.396 | 71.49 | 60.27 | 25.56 | 24.76 |
| Equatorial surface gravity [$m \cdot s^{-2}$] | 3.70 | 8.87 | 9.798 | 3.71 | 24.8 | 10.4 | 8.87 | 11.2 |
| Surface temperature [K] | 440 * | 737 * | 288 * | 210 * | 110 ** | 81 ** | 58 ** | 47 ** |
| Orbital radius *** [$10^9$m] | 57.91 | 108.2 | 149.6 | 227.9 | 778.6 | 1433 | 2873 | 4495 |
| Orbit period [d] | 87.969 | 224.70 | 365.256 | 686.98 | 4332.6 | 10759 | 30685 | 60189 |
| Orbit period [a] | 0.241 | 0.615 | 1.000 | 1.881 | 11.86 | 29.46 | 84.01 | 164.8 |

* Average atmospheric temperature

** Black body temperature

***Semimajor axis

### The Moon

| | |
|---|---|
| Mass | $7.349 \cdot 10^{22}$kg |
| Equatorial radius | $1.738 \cdot 10^6$m |
| Surface gravity | $1.62 m \cdot s^{-2}$ |
| Surface temperature ** | 275K |
| Orbital radius *** | $3.844 \cdot 10^8$m |
| Orbit period | 27.322d |

**Black body temperature

***Semimajor axis

### The Sun

| | |
|---|---|
| Mass | $1.989 \cdot 10^{30}$kg |
| Mean radius | $6.96000 \cdot 10^8$m |
| Surface gravity | $274.0 m \cdot s^{-2}$ |
| Surface temperature ** | 5778K |
| Central temperature | $1.571 \cdot 10^7$K |

**Black body temperature

# 5.  Thermodynamics

## Various thermodynamic data

### *Melting points and boiling points of various substances at standard pressure*

$T_f$: Melting point in K

$\vartheta_f$: Melting point in °C

$T_v$: Boiling point in K

$\vartheta_v$: Boiling point in °C

| Substance | $T_f$ [K] | $T_v$ [K] | $\vartheta_f$ [°C] | $\vartheta_v$ [°C] |
|---|---|---|---|---|
| Antimony | 903.7 | 1713 | 630.5 | 1440 |
| Copper | 1356 | 2840 | 1083 | 2567 |
| Ethyl alcohol | 158.9 | 351.6 | -114.3 | 78.40 |
| Gold | 1336 | 2933 | 1063 | 2660 |
| Helium | 3.550 | 4.216 | -269.6 | -268.9 |
| Hydrogen | 13.84 | 20.26 | -259.3 | -252.9 |
| Lead | 600.5 | 2023 | 327.3 | 1750 |
| Mercury | 234.3 | 629.9 | -38.83 | 356.7 |
| Nitrogen | 63.18 | 77.34 | -210.0 | -195.8 |
| Oxygen | 54.36 | 90.18 | -218.8 | -183.0 |
| Silver | 1234 | 2466 | 960.8 | 2193 |
| Sulfur | 392.0 | 717.8 | 119.0 | 444.6 |
| Water | 273.15 | 373.15 | 0 | 100 |

### Various thermodynamic data of solids at 20°C

$\alpha$: Linear thermal expansion coefficient

$c$: Specific heat capacity

$L_f$: Specific latent heat of fusion

$\lambda$: Heat conduction coefficient

| Solid | $\alpha \left[10^{-6}\mathrm{K}^{-1}\right]$ | $c \left[\mathrm{J} \cdot \mathrm{kg}^{-1} \cdot \mathrm{K}^{-1}\right]$ | $L_f \left[10^{5}\mathrm{J} \cdot \mathrm{kg}^{-1}\right]$ | $\lambda \left[\mathrm{W} \cdot \mathrm{m}^{-1} \cdot \mathrm{K}^{-1}\right]$ |
|---|---|---|---|---|
| Aluminium | 23.7 | 896 | 3.98 | 236 |
| Brass * | 18 | 380 | 1.5 | 120 |
| Cast steel * | 10 | 540 | 1.3 | 50 |
| Copper | 16.7 | 381 | 2.05 | 401 |
| Gold | 14.3 | 130 | 0.63 | 314 |
| Iron | 12.2 | 439 | 2.68 | 80.2 |
| Lead | 29.3 | 129 | 0.25 | 35.3 |
| Platinum | 9.10 | 134 | 1.00 | 71.6 |
| Silver | 19.5 | 235 | 1.05 | 426 |
| Steel * | 13 | 450 | 2.7 | 46 |
| Tungsten | 4.51 | 134 | 1.93 | 167 |
| Zinc | 26.2 | 389 | 1.05 | 110 |
| Ice ** | 37.0 | 2100 | 3.338 | 2.24 |
| Quartz glass * | 0.5 | 710 | | 1.36 |
| Window glass * | 7.6 | 840 | | 1.03 |
| Clay brick * | | 920 | | 0.6 |
| Cork * | | | | 0.04 |
| Oak wood * | 8 | 1600 | | 0.15 |
| Styrofoam * | | 1200 | | 0.03 |

* typical value

** at -4°C

## *Various thermodynamic data of liquids at 20°C*

$\gamma$: Volumetric thermal expansion coefficient

$c$: Specific heat capacity

$L_f$: Specific latent heat of fusion

$L_v$: Specific latent heat of vaporization

$\lambda$: Heat conduction coefficient

| Liquid | $\gamma\,[10^{-3}\mathrm{K}^{-1}]$ | $c\,[10^{3}\mathrm{J}\cdot\mathrm{kg}^{-1}\cdot\mathrm{K}^{-1}]$ | $L_f\,[10^{5}\mathrm{J}\cdot\mathrm{kg}^{-1}]$ | $L_v\,[10^{5}\mathrm{J}\cdot\mathrm{kg}^{-1}]$ | $\lambda\,[\mathrm{W}\cdot\mathrm{m}^{-1}\cdot\mathrm{K}^{-1}]$ |
|---|---|---|---|---|---|
| Acetone | 1.43 | 2.16 | 0.981 | 5.25 | 0.16 |
| Benzene | 1.23 | 1.74 | 1.28 | 3.94 | 0.15 |
| Diethyl ether | 1.62 | 2.31 | 1.00 | 3.84 | 0.13 |
| Ethanol | 1.10 | 2.43 | 1.08 | 8.40 | 0.17 |
| Glycerine (Glycerol) | 0.497 | 2.43 | 2.01 | 8.54 | 0.29 |
| Methanol | 1.10 | 2.47 | 0.921 | 11.0 | 0.20 |
| Mercury | 0.183 | 0.139 | 11.80 | 2.84 | 8.20 |
| Water | 0.208 | 4.18 | 3.34 | 22.57 | 0.60 |

## *Various thermodynamic data of gases at 20°C*

$c_p$: Specific heat capacity if $p$ = const.

$c_V$: Specific heat capacity if $V$ = const.

$\kappa$: Adiabatic coefficient

| Gas | $c_p\,[10^{3}\mathrm{J}\cdot\mathrm{kg}^{-1}\cdot\mathrm{K}^{-1}]$ | $c_v\,[10^{3}\mathrm{J}\cdot\mathrm{kg}^{-1}\cdot\mathrm{K}^{-1}]$ | $\kappa$ |
|---|---|---|---|
| Air | 1.01 | 0.720 | 1.40 |
| Ammonia | 2.19 | 1.66 | 1.32 |
| Argon | 0.519 | 0.318 | 1.64 |
| Carbon dioxide | 0.841 | 0.661 | 1.28 |
| Ethane | 1.75 | 1.48 | 1.18 |
| Helium | 5.19 | 3.13 | 1.65 |
| Hydrogen | 14.3 | 10.2 | 1.41 |
| Methane | 2.22 | 1.67 | 1.32 |
| Natural Gas (typical value) | 2.34 | 1.85 | 1.26 |
| Neon | 1.03 | 0.629 | 1.63 |
| Nitrogen | 1.04 | 0.740 | 1.40 |
| Oxygen | 0.921 | 0.661 | 1.39 |
| Propane | 1.67 | 1.48 | 1.13 |

### Heating values

$H_L$: Heating value

| Substance | $H_L \left[10^6 \text{J} \cdot \text{kg}^{-1}\right]$ ** |
|---|---|
| **Gases and liquids** | |
| Butane | 46 |
| Ethane | 48 |
| Ethanol | 27 |
| Methane | 50 |
| Methanol | 20 |
| Petroleum | 42 |
| Propane | 46 |
| **Fuels and heating fuels** | |
| Biogas * | 25 |
| Coal * | 33 |
| Diesel * | 43 |
| Gasoline * | 43 |
| Hydrogen | 121 |
| Kerosene * | 43 |
| Natural gas * | 39 |
| Paraffin * | 41 |
| **Energy from waste** | |
| Plastic waste * | 30 |
| Biomass * | 17 |
| Residual waste * | 10 |
| **Common materials** | |
| Wood * | 15 |
| Paper * | 13 |
| Polyethylene * | 44 |
| Polypropylene * | 47 |
| Polystyrene * | 42 |

** lower value, i. e. all end products are gaseous

* typical value

### Emissivities at 20°C

$\varepsilon$: Emissivity

| Material | $\varepsilon$ * |
|---|---|
| Asphalt | 0.93 |
| Black body | 1 |
| Carbon | 0.8 |
| Clay brick | 0.75 |
| Concrete | 0.85 |
| Cotton cloth | 0.77 |
| Glass | 0.9 |
| Granite | 0.45 |
| Ice | 0.96 |
| Iron (oxidized) | 0.5 |
| Iron (polished) | 0.04 |
| Iron (rough) | 0.25 |
| Paint | 0.96 |
| Sand | 0.75 |
| Skin | 0.97 |
| Textiles | 0.85 |
| Water | 0.98 |
| Wood (planed) | 0.9 |

* typical values, depending on exact surface properties

## Heat transfer coefficients

$U$: Heat transfer coefficient

|  | Building element | Thickness $d$ $[10^{-2}\text{m}]$ | $U$ $[\text{W} \cdot \text{m}^{-2} \cdot \text{K}^{-1}]$ * |
|---|---|---|---|
| **External walls** | Concrete | 25 | 3.3 |
|  | Clay brick | 25 | 1.5 |
|  | Clay brick | 37 | 0.8 |
|  | Highly porous clay brick | 50 | 0.19 |
|  | Solid wood | 20 | 0.5 |
|  | Cellular concrete | 37 | 0.2 |
|  | Cellular concrete | 40 | 0.18 |
|  | Cellular concrete | 50 | 0.14 |
| **Internal walls** | Clay brick | 12 | 3 |
|  | Wood | 5 | 2.6 |
|  | Cellular concrete | 28 | 0.6 |
| **Doors** | Wood or plastic | - | 3.5 |
| **Windows** | Single glazed | 0.4 | 5.9 |
|  | Double glazed | 2.5 | 2.9 |
|  | Triple glazed | 2.5 | 1.3 |
|  | Passive house window | - | 0.6 |

* typical values, depending on the exact properties of the building element

## 6.  Electrodynamics

### Resistivities at 20°C

$\rho_{el}$: Resistivity

| Substance | $\rho_{el}\ [\Omega \cdot \mathrm{m}]$ |
|---|---|
| **Conductors** | |
| Aluminium | $2.8 \cdot 10^{-8}$ |
| Brass (typical value) | $7 \cdot 10^{-8}$ |
| Constantan | $4.9 \cdot 10^{-7}$ |
| Copper | $1.7 \cdot 10^{-8}$ |
| Gold | $2.3 \cdot 10^{-8}$ |
| Iron | $9.8 \cdot 10^{-8}$ |
| Lead | $2.2 \cdot 10^{-7}$ |
| Nickel | $7.8 \cdot 10^{-8}$ |
| Platinum | $1.1 \cdot 10^{-7}$ |
| Silver | $1.6 \cdot 10^{-8}$ |
| Zinc | $5.8 \cdot 10^{-8}$ |

| Substance | $\rho_{el}\ [\Omega \cdot \mathrm{m}]$ |
|---|---|
| **Semiconductors** | |
| Germanium (pure) | 0.46 |
| Silicon (pure) | $2.3 \cdot 10^{3}$ |
| **Insulators (typical values)** | |
| Glass | $10^{12}$ |
| Hard rubber | $10^{14}$ |
| Paraffin | $10^{17}$ |
| PET | $10^{20}$ |
| Quartz glass | $10^{15}$ |
| Teflon | $10^{18}$ |

### Relative permeabilities at 20°C

$\mu_r$: Relative permeability

| Substance | $\mu_r$ |
|---|---|
| Vacuum | 1 |
| Superconductor | 0 |
| **Paramagnetic substances** | |
| Air (at 3 bar) | 1.00000037 |
| Aluminium | 1.00002 |
| Chromium | 1.00029 |
| Oxygen | 1.0000018 |
| Platinum | 1.00026 |

| Substance | $\mu_r$ |
|---|---|
| **Diamagnetic substances** | |
| Copper | 0.99999 |
| Lead | 0.99984 |
| Mercury | 0.99997 |
| Water | 0.999991 |
| **Ferromagnetic substances *** | |
| Iron | 7000 |
| Nickel | 900 |
| Cobalt | 250 |
| $\mu$ -metal | 10000 |
| Permalloy | 300000 |
| Special alloys | 800000 |

* maximum value

## Relative permittivities at 20°C

$\varepsilon_r$: Relative permittivity

| | Substance | $\varepsilon_r$ ** |
|---|---|---|
| | Vacuum | 1 |
| **Gases *** | Air (standard pressure) | 1.0006 |
| | Hydrogen | 1.00026 |
| | Oxygen | 1.00055 |
| **Liquids** | Benzene | 2.3 |
| | Ethanol | 25 |
| | Glycerine (Glycerol) | 41 |
| | Petroleum | 2.1 |
| | Water | 80 |
| **Solids** | Amber | 2.7 |
| | Ceramics ($Ca\ TiO_3$) | 160 |
| | Ceramics ($TiO_2$) | 80 |
| | Germanium | 16 |
| | Glass | 6.0 |
| | Hard rubber * | 3.5 |
| | Mica * | 6.5 |
| | Plexiglass * | 3.4 |
| | Porcelain * | 6.0 |
| | Quartz glass * | 4.2 |
| | Silicon | 12 |
| | Wood * | 5 |

* typical value

** static, i. e. in a constant electrical field

*** at standard pressure

# 7.  Optics

## Spectrum of light (with Fraunhofer lines  of the Sun)

$\lambda$: Wavelength

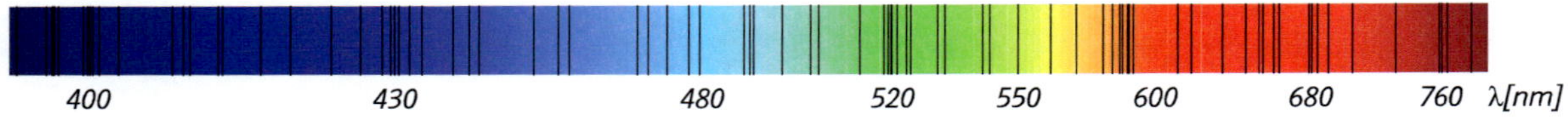


## Refractive indices

$n$: Refractive index

| Substance | $n$ | | Substance | $n$ |
|---|---|---|---|---|
| Vacuum | 1 | | **Gases at 0 °C** | |
| | | | Air | 1.00029 |
| **Solids at 20°C** | | | Carbon dioxide | 1.00045 |
| Acetone | 1.36 | | Helium | 1.00004 |
| Amber | 1.55 | | Hydrogen | 1.00013 |
| Crown glass (pure) | 1.52 | | | |
| Diamond | 2.42 | | **Liquids at 20 °C** | |
| Flint glass (pure) | 1.61 | | Benzene | 1.5 |
| Ice | 1.31 | | Cedar oil | 1.5 |
| Jena glass SFS1 | 1.92 | | Ethanol | 1.36 |
| PET | 1.58 | | Silicone oil | 1.52 |
| Plexiglass | 1.49 | | Water | 1.33 |
| Sapphire | 1.77 | | | |
| Silicon | 4.01 | | **Other materials** | |
| Table salt | 1.5 | | Cornea (eye) | 1.34 |
| | | | Eye lens | 1.39 |

## 8. Radioactivity

### Radioactive isotopes and decays

$T_{1/2}$: Half-life

| Isotope | Full name | Decay * | $T_{1/2}$ |
|---|---|---|---|
| Ac-228 | Actinium-228 | $\beta$ | 6.13h |
| Bi-210 | Bismuth-210 | $\beta$ | 5d |
| Bi-211 | Bismuth-211 | $\alpha$ | 2.13min |
| Bi-214 | Bismuth-214 | $\beta$ | 19.9min |
| C-14 | Carbon-14 | $\beta$ | 5736a |
| Pb-210 | Lead-210 | $\beta$ | 22.3a |
| Pb-211 | Lead-211 | $\beta$ | 36.2min |
| Pb-212 | Lead-212 | $\beta$ | 10.63h |
| Pb-214 | Lead-214 | $\beta$ | 26.9min |
| Po-210 | Polonium-210 | $\alpha$ | 139d |
| Po-214 | Polonium-214 | $\alpha$ | 164µs |
| Po-215 | Polonium-215 | $\alpha$ | 1.78ms |
| Po-216 | Polonium-216 | $\alpha$ | 150ms |
| Po-218 | Polonium-218 | $\alpha$ | 3.05min |
| Pa-231 | Protactinium-231 | $\alpha$ | 32760a |
| Pa-234 | Protactinium-234 | $\beta$ | 1.19min |
| Ra-223 | Radium-223 | $\alpha$ | 11.4d |
| Ra-224 | Radium-224 | $\alpha$ | 3.65d |
| Ra-226 | Radium-226 | $\alpha$ | 1602a |
| Rn-219 | Radon-219 | $\alpha$ | 3.97s |
| Rn-220 | Radon-220 | $\alpha$ | 55.5s |
| Rn-222 | Radon-222 | $\alpha$ | 3.82d |
| Tl-207 | Thallium-207 | $\beta$ | 4.8min |
| Th-228 | Thorium-228 | $\alpha$ | 1.91a |
| Th-230 | Thorium-230 | $\alpha$ | 75380a |
| Th-231 | Thorium-231 | $\beta$ | 25.7h |
| Th-232 | Thorium-232 | $\alpha$ | $14.05 \cdot 10^9$a |
| Th-234 | Thorium-234 | $\beta$ | 24.2d |
| U-234 | Uranium-234 | $\alpha$ | $246 \cdot 10^3$a |
| U-235 | Uranium-235 | $\alpha$ | $704 \cdot 10^6$a |
| U-238 | Uranium-238 | $\alpha$ | $4.47 \cdot 10^9$a |

* main decay mode

## Radioactive decay chains

$T_{1/2}$: Half-life

### Uranium-Radium

| Nuclide | $T_{1/2}$ | Decay mode | Decay product |
|---|---|---|---|
| U-238 | $4.468 \cdot 10^9$ a | $\alpha$ | Th-234 |
| Th-234 | 24.10 d | $\beta$ | Pa-234 |
| Pa-234 | 6.7 h | $\beta$ | U-234 |
| U-234 | $2.455 \, 10^5$ a | $\alpha$ | Th-230 |
| Th-230 | $7.538 \, 10^4$ a | $\alpha$ | Ra-226 |
| Ra-226 | 1602 a | $\alpha$ | Rn-222 |
| Rn-222 | 3.8235 d | $\alpha$ | Po-218 |
| Po-218 | 3.05 min | $\alpha$ | Pb-214 |
| Pb-214 | 26.8 min | $\beta$ | Bi-214 |
| Bi-214 | 19.9 min | $\beta$ | Po-214 |
| Po-214 | 164 μs | $\alpha$ | Pb-210 |
| Pb-210 | 22.3 a | $\beta$ | Bi-210 |
| Bi-210 | 5.013 d | $\beta$ | Po-210 |
| Po-210 | 138.376 d | $\alpha$ | Pb-206 |
| Pb-206 | stable | - | - |

### Neptunium

| Nuclide | $T_{1/2}$ | Decay mode | Decay product |
|---|---|---|---|
| Pu-241 | 14.4 a | $\beta$ | Am-241 |
| Am-241 | 432.2 a | $\alpha$ | Np-237 |
| Np-237 | $2.144 \cdot 10^6$ a | $\alpha$ | Pa-233 |
| Pa-233 | 26.967 d | $\beta$ | U-233 |
| U-233 | $1.592 \cdot 10^5$ a | $\alpha$ | Th-229 |
| Th-229 | 7880 a | $\alpha$ | Ra-225 |
| Ra-225 | 14.9 d | $\beta$ | Ac-225 |
| Ac-225 | 10.0 d | $\alpha$ | Fr-221 |
| Fr-221 | 4.9 min | $\alpha$ | At-217 |
| At-217 | 32.3 ms | $\alpha$ | Bi-213 |
| Bi-213 | 45.49 min | $\beta$ (97.91%) | Po-213 |
| Bi-213 | 45.49 min | $\alpha$ (2.09%) | Tl-209 |
| Po-213 | 4.2 ms | $\alpha$ | Pb-209 |
| Tl-209 | 2.20 min | $\beta$ | Pb-209 |
| Pb-209 | 3.253 h | $\beta$ | Bi-209 |
| Bi-209 | $1.9 \cdot 10^{19}$ a | $\alpha$ | Tl-205 |
| Tl-205 | stable | - | - |

## Thorium

| Nuclide | $T_{1/2}$ | Decay mode | Decay product |
|---|---|---|---|
| Th-232 | $1.405 \cdot 10^{10}$ a | $\alpha$ | Ra-228 |
| Ra-228 | 5.75 a | $\beta$ | Ac-228 |
| Ac-228 | 6.15 h | $\beta$ | Th-228 |
| Th-228 | 1.9131 a | $\alpha$ | Ra-224 |
| Ra-224 | 3.66 d | $\alpha$ | Rn-220 |
| Rn-220 | 55.6 s | $\alpha$ | Po-216 |
| Po-216 | 0.145 s | $\alpha$ | Pb-212 |
| Pb-212 | 10.64 h | $\beta$ | Bi-212 |
| Bi-212 | 60.55 min | $\beta$ (64.06%) | Po-212 |
| | | $\alpha$ (35.94%) | Tl-208 |
| Po-212 | $2.99 \cdot 10^{-7}$ s | $\alpha$ | Pb-208 |
| Tl-208 | 3.083 min | $\beta$ | Pb-208 |
| Pb-208 | stable | - | - |

## Uranium-Actinium

| Nuclide | $T_{1/2}$ | Decay mode | Decay product |
|---|---|---|---|
| Pu-239 | $2.411 \cdot 10^{4}$ a | $\alpha$ | U-235 |
| U-235 | $7.038 \cdot 10^{8}$ a | $\alpha$ | Th-231 |
| Th-231 | 25.52 h | $\beta$ | Pa-231 |
| Pa-231 | $3.276 \cdot 10^{4}$ a | $\alpha$ | Ac-227 |
| Ac-227 | 21.773 a | $\beta$ (98.62%) | Th-227 |
| | | $\alpha$ (1.38%) | Fr-223 |
| Th-227 | 18.72 d | $\alpha$ | Ra-223 |
| Fr-223 | 22.0 min | $\beta$ | Ra-223 |
| Ra-223 | 11.435 d | $\alpha$ | Rn-219 |
| Rn-219 | 3.96 s | $\alpha$ | Po-215 |
| Po-215 | 1.781 ms | $\alpha$ | Pb-211 |
| Pb-211 | 36.1 min | $\beta$ | Bi-211 |
| Bi-211 | 2.14 min | $\alpha$ | Tl-207 |
| Tl-207 | 4.77 min | $\beta$ | Pb-207 |
| Pb-207 | stable | - | - |

## 9. Chemistry

*Periodic table of the elements*

| Period \ Group | IA | IIA | IIIB | IVB | VB | VIB | VIIB | VIII | VIII | VIII | IB | IIB | IIIA | IVA | VA | VIA | VIIA | VIIIA |
|---|---|---|---|---|---|---|---|---|---|---|---|---|---|---|---|---|---|---|
| 1 | 1 H 1.008 | | | | | | | | | | | | | | | | | 2 He 4.003 |
| 2 | 3 Li 6.941 | 4 Be 9.012 | | | | | | | | | | | 5 B 10.81 | 6 C 12.01 | 7 N 14.01 | 8 O 16 | 9 F 19 | 10 Ne 20.18 |
| 3 | 11 Na 22.99 | 12 Mg 24.31 | | | | | | | | | | | 13 Al 26.98 | 14 Si 28.09 | 15 P 30.97 | 16 S 32.07 | 17 Cl 35.45 | 18 Ar 39.95 |
| 4 | 19 K 39.1 | 20 Ca 40.08 | 21 Sc 44.96 | 22 Ti 47.88 | 23 V 50.94 | 24 Cr 52 | 25 Mn 54.94 | 26 Fe 55.85 | 27 Co 58.93 | 28 Ni 58.69 | 29 Cu 63.55 | 30 Zn 65.39 | 31 Ga 69.72 | 32 Ge 72.59 | 33 As 74.92 | 34 Se 78.96 | 35 Br 79.9 | 36 Kr 83.8 |
| 5 | 37 Rb 85.47 | 38 Sr 87.62 | 39 Y 88.91 | 40 Zr 91.22 | 41 Nb 92.91 | 42 Mo 95.94 | 43 Tc [98] | 44 Ru 101.1 | 45 Rh 102.9 | 46 Pd 106.4 | 47 Ag 107.9 | 48 Cd 112.4 | 49 In 114.8 | 50 Sn 118.7 | 51 Sb 121.8 | 52 Te 127.6 | 53 I 126.9 | 54 Xe 131.3 |
| 6 | 55 Cs 132.9 | 56 Ba 137.3 | 57 La 138.9 | 72 Hf 178.5 | 73 Ta 180.9 | 74 W 183.9 | 75 Re 186.2 | 76 Os 190.2 | 77 Ir 192.2 | 78 Pt 195.1 | 79 Au 197 | 80 Hg 200.5 | 81 Tl 204.4 | 82 Pb 207.2 | 83 Bi 209 | 84 Po [210] | 85 At [210] | 86 Rn [222] |
| 7 | 87 Fr [223] | 88 Ra [226] | 89 Ac [227] | 104 Rf [257] | 105 Db [260] | 106 Sg [263] | 107 Bh [262] | 108 Hs [265] | 109 Mt [266] | 110 Ds [269] | 111 Rg [272] | 112 Cn [277] | 113 Uut [?] | 114 Uuq [285] | 115 Uup [?] | 116 Uuh [289] | 117 Uus [?] | 118 Uuo [?] |

**Lanthanide Series**

| 58 Ce 140.1 | 59 Pr 140.9 | 60 Nd 144.2 | 61 Pm [147] | 62 Sm 150.4 | 63 Eu 152 | 64 Gd 157.3 | 65 Tb 158.9 | 66 Dy 162.5 | 67 Ho 164.9 | 68 Er 167.3 | 69 Tm 168.9 | 70 Yb 173 | 71 Lu 175 |
|---|---|---|---|---|---|---|---|---|---|---|---|---|---|

**Actinide Series**

| 90 Th 232 | 91 Pa [231] | 92 U [238] | 93 Np [237] | 94 Pu [242] | 95 Am [243] | 96 Cm [247] | 97 Bk [247] | 98 Cf [249] | 99 Es [254] | 100 Fm [253] | 101 Md [256] | 102 No [254] | 103 Lr [257] |
|---|---|---|---|---|---|---|---|---|---|---|---|---|---|

The table cells indicate the atomic number, the atomic symbol and the average (atomic) mass number of each element. Average mass numbers in parentheses indicate the mass number of the longest-living isotope.

# IV.  Mathematics

## 1.  Algebra

**Exponentiation: Powers of ten**

| | |
|---|---|
| $10^n = 1\underbrace{0\ldots0}_{n}$ | (A "one with n zeros") |
| $10^{-n} = 0.\underbrace{0\ldots1}_{n}$ | (The nth decimal place is 1.) |
| $10^0 = 1$ | |
| $10^{-n} = \dfrac{1}{10^n}$ | |
| $10^n = \dfrac{1}{10^{-n}}$ | |
| $10^n \cdot 10^m = 10^{n+m}$ | |
| $(10^n)^m = 10^{n\cdot m}$ | |

## 2.  Analysis

**Real functions**

A real function is a mapping

$$y: \begin{cases} D \to \mathbb{R} \; where \; D \subset \mathbb{R} \\ x \to y(x) \end{cases}$$

**Real functions: The linear function**

A real function

$$y: \begin{cases} \mathbb{R} \to \mathbb{R} \\ x \to a \cdot x + b \; where \; a, b \in \mathbb{R} \end{cases}$$

is a linear function.

Its graph is a straight line.

$$a = \frac{\Delta y}{\Delta x} = \frac{y_2 - y_1}{x_2 - x_1} \text{ is its slope.}$$

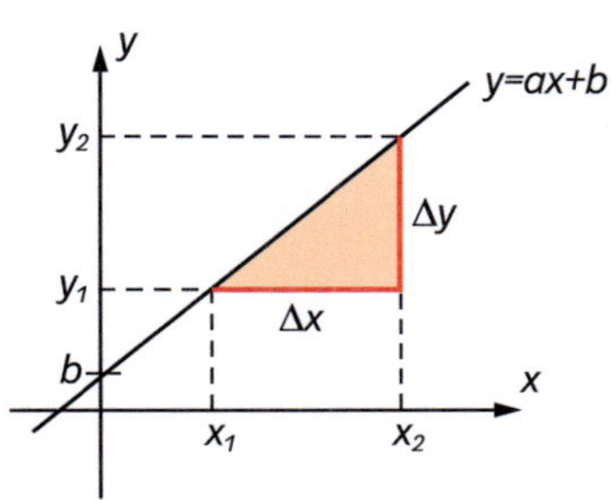


$b = y(0)$ is its y-intercept.

## Real functions: The straight line through zero

A linear function with b=0

$$y: \begin{cases} \mathbb{R} \to \mathbb{R} \\ x \to a \cdot x \ where \ a, b \in \mathbb{R} \end{cases}$$

is called a straight line through zero.

$$y = a \cdot x$$

y is proportional to x: $y \sim x$

a is the constant of proportionality.

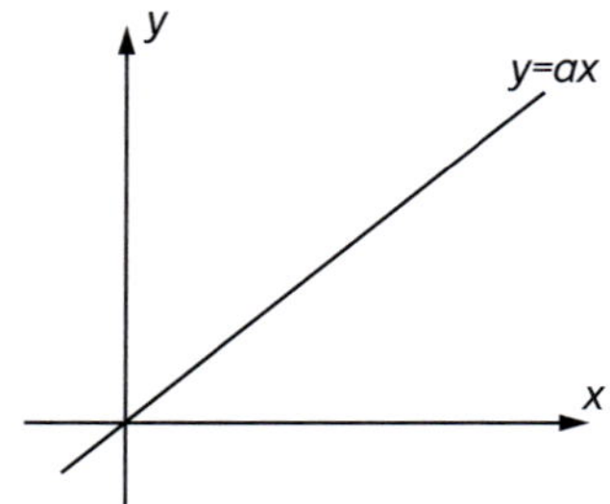


## Vector analysis: Vectors

A vector $\vec{a}$ has a magnitude (or norm) $a = |\vec{a}|$ and a direction. If represented by an arrow, its magnitude corresponds to the arrow's length.

## Vector analysis: Resolving vectors

$$\vec{a} = \vec{a}_x + \vec{a}_y$$

$$a^2 = a_x^2 + a_y^2$$

$$a_y = a \cdot \sin \alpha$$

$$a_x = a \cdot \cos \alpha$$

$$\tan \alpha = \frac{a_y}{a_x}$$

$$\tan \beta = \frac{a_x}{a_y}$$

$$\sin \alpha = \cos \beta$$

$$\sin \beta = \cos \alpha$$

$$\alpha + \beta = 90°$$

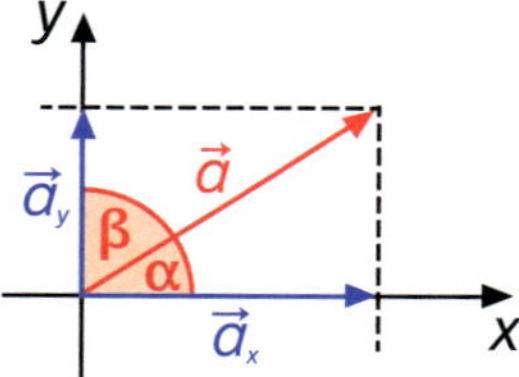

# 3. Geometry

$P$: Perimeter

$A$: Area resp. Surface area

$V$: Volume

## Properties of geometric objects

| | | |
|---|---|---|
| Square | 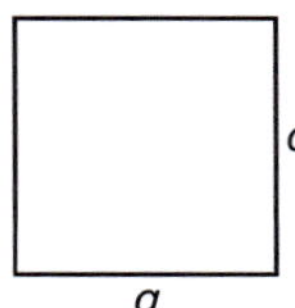 | $P = 4 \cdot a$<br>$A = a^2$ |
| Rectangle | 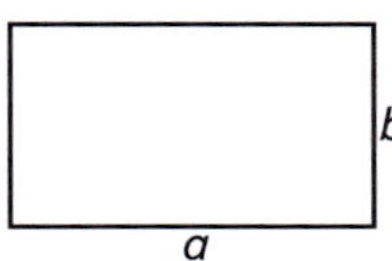 | $P = 2 \cdot a + 2 \cdot b = 2 \cdot (a + b)$<br>$A = a \cdot b$ |
| Triangle | 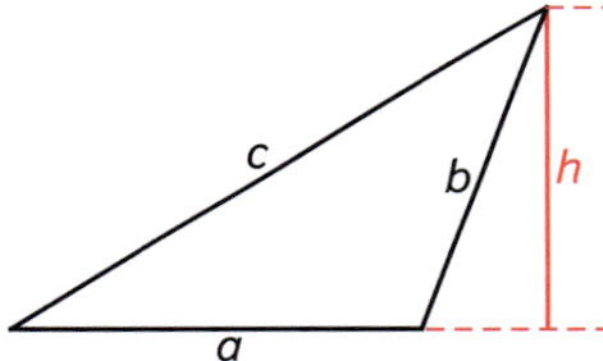 | $P = a + b + c$<br>$A = \dfrac{1}{2} \cdot a \cdot h$ |
| Parallelogram | 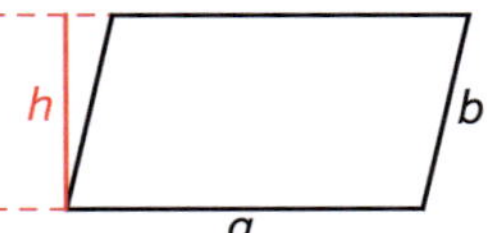 | $P = 2 \cdot a + 2 \cdot b = 2 \cdot (a + b)$<br>$A = a \cdot h$ |
| Trapezoid | 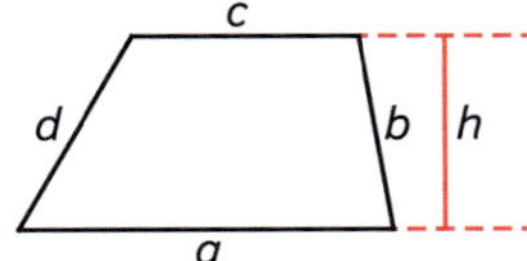 | $P = a + b + c + d$<br>$A = \dfrac{a + c}{2} \cdot h$ |
| Circle | 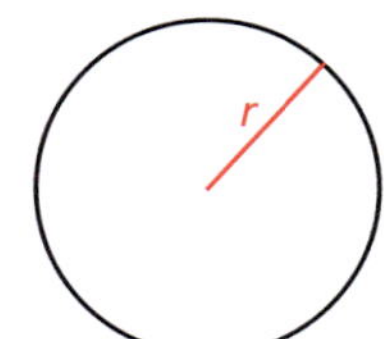 | Circumference $C = 2 \cdot \pi \cdot r$<br>$A = \pi \cdot r^2$ |

| | | |
|---|---|---|
| Cube | 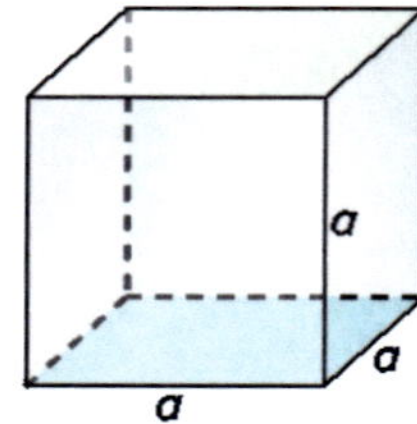 | $A = 6 \cdot a^2$ <br> $V = a^3$ |
| Cuboid | 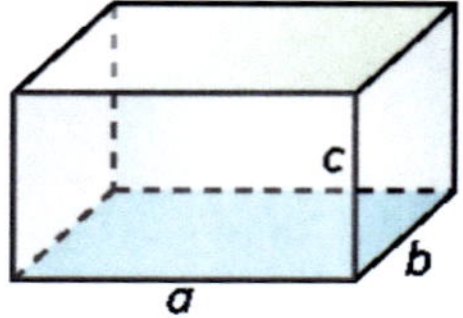 | $A = 2 \cdot (a \cdot b + b \cdot c + a \cdot c)$ <br> $V = a \cdot b \cdot c$ |
| Sphere | 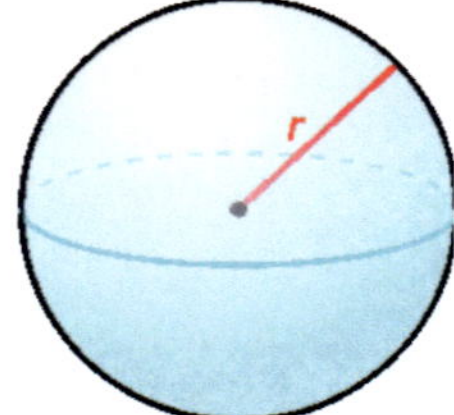 | $A = 4 \cdot \pi \cdot r^2$ <br> $V = \dfrac{4}{3} \cdot \pi \cdot r^3$ |
| Cylinder | 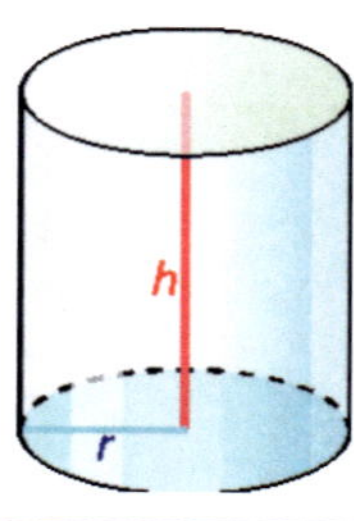 | $A = 2 \cdot \pi \cdot r^2 + 2 \cdot \pi \cdot r \cdot h = 2 \cdot \pi \cdot r \cdot (r + h)$ <br> $V = \pi \cdot r^2 \cdot h$ |

## Trigonometry: The right triangle

$$\alpha + \beta = 90°$$

$$a^2 + b^2 = c^2$$

$$\sin \alpha = \frac{a}{c} = \frac{opposite}{hypotenuse}$$

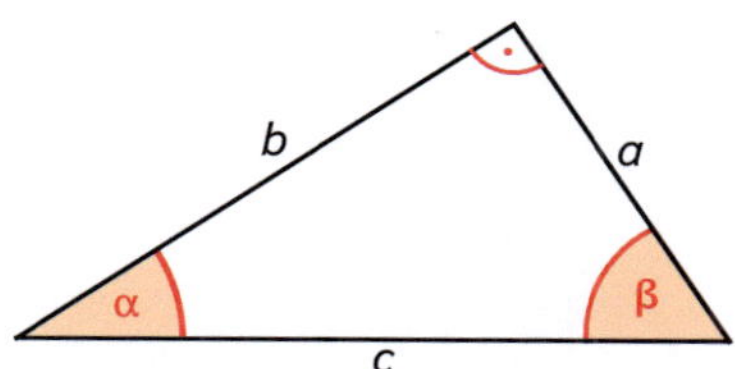

$$\alpha = \arcsin \left(\frac{a}{c}\right)$$

$$\cos \alpha = \frac{b}{c} = \frac{adjacent}{hypotenuse}$$

$$\alpha = \arccos \left(\frac{b}{c}\right)$$

$$\tan \alpha = \frac{a}{b} = \frac{opposite}{adjacent} = \frac{\sin \alpha}{\cos \alpha}$$

$$\alpha = \arctan \left(\frac{a}{b}\right)$$

$$\sin \alpha = \cos \beta$$

$$\sin \beta = \cos \alpha$$

$$A = \frac{a \cdot b}{2}$$

## Trigonometry: The triangle

$$\alpha + \beta + \gamma = 180°$$

$$a^2 = b^2 + c^2 - 2bc \cdot \cos \alpha$$

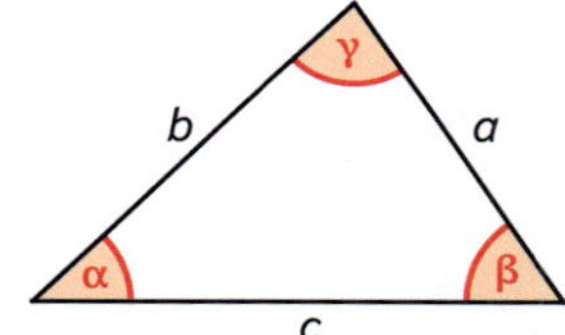

$$\frac{\sin \alpha}{a} = \frac{\sin \beta}{b} = \frac{\sin \gamma}{c}$$

$$A = \frac{a \cdot b}{2} \cdot \sin \gamma$$

# 4. Statistics and error

## Statistics

| | |
|---|---|
| Arithmetic mean | $\displaystyle \bar{x} = \frac{1}{n} \sum_{i=1}^{n} x_i = \frac{1}{n}(x_1 + x_2 + \cdots + x_n)$ |
| Variance | $\displaystyle \sigma^2 = \frac{1}{n} \sum_{i=1}^{n} (x_i - \bar{x})^2 = \frac{1}{n} \cdot [(x_1 - \bar{x})^2 + (x_2 - \bar{x})^2 + \cdots + (x_n - \bar{x})^2]$ |
| Standard deviation | $\displaystyle \sigma = \sqrt{\frac{1}{n} \sum_{i=1}^{n} (x_i - \bar{x})^2} = \sqrt{\frac{1}{n} \cdot [(x_1 - \bar{x})^2 + (x_2 - \bar{x})^2 + \cdots + (x_n - \bar{x})^2]}$ |

## Error and error propagation

| | |
|---|---|
| Absolute error on a quantity x | $\Delta x$ |
| Relative error on a quantity x | $\dfrac{\Delta x}{x}$ |
| Mean error (n measurements of the same measurand) | $\displaystyle \overline{\Delta x} = \frac{1}{n} \sum_{i=1}^{n} |x_i - \bar{x}| = \frac{1}{n} \cdot (|x_1 - \bar{x}| + |x_2 - \bar{x}| + \cdots + |x_n - \bar{x}|)$ |
| Sum or difference: absolute error | $a = x + y - z \Rightarrow \Delta a = \sqrt{(\Delta x)^2 + (\Delta y)^2 + (\Delta z)^2}$ |
| Sum or difference: maximum absolute error | $a = x + y - z \Rightarrow \Delta a \leq \Delta x + \Delta y + \Delta z$ |
| Sum or difference: relative error | $a = x + y - z \Rightarrow \dfrac{\Delta a}{|a|} = \dfrac{\sqrt{(\Delta x)^2 + (\Delta y)^2 + (\Delta z)^2}}{|a|}$ |
| Product or quotient: absolute error | $a = \dfrac{x \cdot z}{y} \Rightarrow \Delta a = |a| \cdot \sqrt{\left(\dfrac{\Delta x}{x}\right)^2 + \left(\dfrac{\Delta y}{y}\right)^2 + \left(\dfrac{\Delta z}{z}\right)^2}$ |
| Product or quotient: relative error | $a = \dfrac{x \cdot z}{y} \Rightarrow \dfrac{\Delta a}{|a|} = \sqrt{\left(\dfrac{\Delta x}{x}\right)^2 + \left(\dfrac{\Delta y}{y}\right)^2 + \left(\dfrac{\Delta z}{z}\right)^2}$ |

# V.    Index